I0797680

Tiny Gardens Everywhere

ALSO BY KATE BROWN

Manual for Survival: A Chernobyl Guide to the Future

Plutopia: Nuclear Families, Atomic Cities, and the Great Soviet and American Plutonium Disasters

Dispatches from Dystopia: Histories of Places Not Yet Forgotten

A Biography of No Place: From Ethnic Borderland to Ukrainian Heartland

Tiny Gardens Everywhere

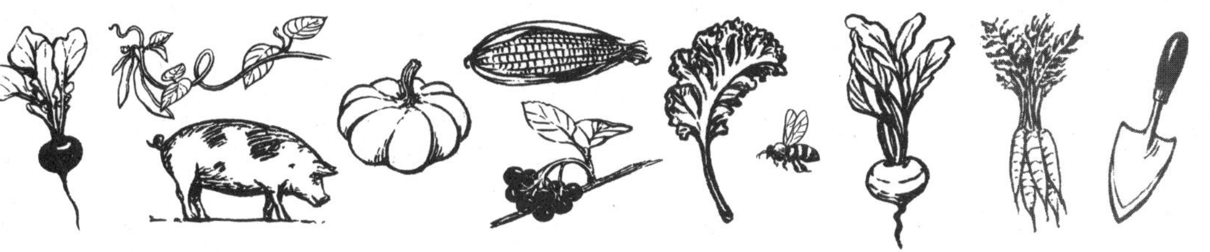

The Past, Present, and Future of the Self-Provisioning City

Kate Brown

W. W. NORTON & COMPANY
Independent Publishers Since 1923

Tiny Gardens Everywhere is a work of nonfiction. Certain names have been changed, as have potentially identifying characteristics.

Printed in the United States of America
First Edition

For information about special discounts for bulk purchases, please contact W. W. Norton Special Sales at specialsales@wwnorton.com or 800-233-4830

Manufacturing by Lake Book Manufacturing
Book design by Dana Sloan
Production manager: Louise Mattarelliano

Library of Congress Cataloging-in-Publication Data is available.

ISBN 978-1-324-10583-1

W. W. Norton & Company, Inc., 500 Fifth Avenue, New York, NY 10110
www.wwnorton.com
W. W. Norton & Company Ltd., 15 Carlisle Street, London W1D 3BS

Authorized EU representative: EAS, Mustamäe tee 50, 10621 Tallinn, Estonia

10 9 8 7 6 5 4 3 2 1

To my parents, Bill and Sally Brown, two people dedicated to caring (for four children and eight grandchildren, for neighbors and extended family, for the wayward at the doorstep, for plants, bees, dogs, cats, and any little spider trapped in the bathtub)

Contents

Tiny Gardens Everywhere

A thriving allotment garden in a bomb crater. London, 1943.

Introduction

BEES HUMMING. MUSHROOMS SPROUTING. Gardens of flowers and vegetables. That's not what we usually associate with urban neighborhoods from the nightly news. But researchers are finding that more raccoons, foxes, and coyotes live in American cities than they do in the countryside. New York City has the highest density of the once-endangered peregrine falcon in the world. Where I live in Cambridge, Massachusetts, wild turkeys stroll down the sidewalks, amicably making way for the human pedestrians. Urban animals and birds rely on a mosaic of small green spaces for their existence. How did that happen? How did cities become refuges for plant and animal life?

The usual story historians tell about urbanization is that in the nineteenth and twentieth centuries, masses flocked to cities looking for industrial jobs. Millions of people overtaxed scarce resources, and soon cities were filthy, smoggy, hungry, disease-ridden places. Urban planners fretted over "asphalt jungles," full of rodents and criminals. Over time, city leaders paved the ground, housed people in ever-taller buildings, and applied weed killer, as if removing animal, vegetable, and mineral matter from cities would also tame the wild and uncivilized working classes. Sewer systems and pavement saved urban residents from epidemics, but in the twentieth century, those who had the choice to moved away to cleaner, greener suburbs. In the following decades, agriculture became more technologically sophisticated, and fewer farmers were needed on vast stretches of land to grow food for billions of people on earth, most of them urbanized. Granted, the food was less nutritious, and the land it was grown on was troubled with toxins, but the fact that the "population bomb" did not detonate, the story goes, was due to a Green Revolution in agriculture in the 1950s, which profited from new hybrid seeds, heavy machinery, fertilizers, and pesticides that increased agricultural yields.

Yet, for all the pages devoted to it, these versions of urban and agricultural history do not explain how, as cities turned into brutal assemblies of concrete, high rises, and whizzing machines, they also became hotspots of biodiversity, nor do they account for why experts worry today about both the precarity of industrial agriculture and a coming "polyfood crisis."

There is another history I would like to tell in these pages, one that has been overlooked in plain sight. Urban gardeners and farmers in cities in Europe and North America throughout the long twentieth century accomplished many of the goals of contemporary food sustainability reformers. With no tax breaks, regulatory structure, or founding manifestos, urban growers quietly produced local, diverse, organic fruits and vegetables on marginal land with short market chains in a production cycle that resulted in affordable, fresh, and nutritious food. They did not expend much in the way of fossil fuels to till, fertilize, harvest, package, or ship produce. Urban gardeners repurposed mountains of garbage into nutrients that saved them from wearing out their soils and needing to go on crusades to colonize other people's land. Tiny gardens stitched together urban ecosystems—one small garden, one planted embankment, one allotment at a time—to fashion metropolitan fabrics that have become today some of the most vital sanctuaries for plants, birds, insects, animals, and humans.

I am an environmental historian. I normally write about modern wastelands and technological disasters, but for the past five years I have taken an interest in how plants in their steady labor of photosynthesis build infrastructure and nurture communities. Following my curiosity, I went to archives, as historians do, but I also started projects to plant urban food landscapes and joined ongoing endeavors in Washington, DC; Cambridge, Massachusetts; Mansfield, Ohio; and Amsterdam, the Netherlands. The same research could have been done in countless cities, but, for the most part, I stuck to places where I happened to be living or visiting anyway.

A lifelong gardener, I spent many weekends of my childhood in my mother's garden in the backyard or on my grandparents' farm near Chicago. I grasped the difference between the hard work of farming and the easy pleasures of gardening. I knew that good gardeners allow the mysterious actions of soil, fungi, and plants to do most of the heavy lifting of growing food. In the decades that followed working my own small plots in urban community gardens, I discovered that a person can feed a family for most months of the year with produce from a few dozen square meters. And I found out why that was so, why the most fertile agriculture in recorded human history occurred not in farm fields (the product of gigantic exertions of fossil-fuel energy, science, and technology) but with little effort in small garden beds. I also learned that when people come together to grow food, they cultivate other interests in economics and politics, in equity and justice, and in healthy bodies and ecologies. The gardeners acted in response to massive changes occurring around them.

In 1850 or soon after, cotton, wheat, sugar, coal, oil, gas, timber, steel, and other commodities began to pour through the gates of European and North American cities. Bodies followed the materials. Bodies of artisans and builders, of farmers who could not make rent, of peasants who could no longer glean from rural commons and wastes once they had been privatized. Rural migrants brought to cities small and large animals. They carried seeds and plants. Wild animals, with less to eat in a shrinking wilderness, also found their way to urban territory. Pigeons, gulls, rats, possums, skunks, songbirds, and waterfowl were drawn to cities for the same reason that millions of working poor trundled into urban areas in the twentieth century. They were all fleeing scarcity in depleted rural landscapes.

Once they landed in cities, working people set about re-creating what they knew. They found urban wastes and made use of them. They came together to rebuild commons, asserting their rights, not to abstractions such as liberty, freedom, and the pursuit of happiness (who can eat that?) but common-law rights to food, fuel, and shelter that they had lost upon enclosures of common land in their villages. Working people planted wild gardens without permission or guidance from authorities—a radical act. They wandered to urban margins to forage for wild greens, berries, and mushrooms, and to gather wood for fencing and fuel. On these edges, they found refuge from noisy and crowded tenements and slums. They protected these wild spaces from encroachment. Thanks in part to the poor people who cared for and defended urban green space, we have Central Park in New York, Hampstead Heath in London, Berlin's encircling green belt, and Rock Creek Park in Washington, DC.

Working people built tiny gardens wherever they could to feed themselves spiritually and physically. For fertilizer, they made use of great stores of organic nutrients that wash up in cities—food scraps, ash, manure, anything that rots fruitfully. Using a river of organic materials flowing into cities, they reengineered sand, swamp, and stone into a good, dark tilth. They erected orchards along warm, south-facing urban walls. They populated courtyards, alleys, and vacant lots with berries and climbing squash. Often working together in collectives innocuously called "garden associations," they built radically egalitarian settlements long before the English socialist, Ebenezer Howard, came up with the concept of "garden cities."

They are still at it, I found, in a rustbelt Midwestern city, on the edges of Amsterdam, and in the southeast quadrant of Washington, DC, where I met a young gardener by the name of J.J. Boone.

He carried a small willow oak log drilled with one-inch holes to inoculate with the spawn of oyster mushrooms. J.J.'s hands didn't look like a gardener's. He had long delicate fingers with neatly trimmed nails. He pointed to the sawdust packed around the fungi. "That's their appetizer, and when we put them in this log, it becomes their entrée."

J.J., about twenty-five, was there to tell me and a group of volunteers about beekeeping. His hands drew the tails of worker bees in the air. As he spoke, J.J. shifted around, his tall form mimicking the shape of bee movements.

"I finally realized after working with bees for a long time that we human beekeepers don't do anything. We just open up the hive and make sure there's no problem. The bees do all the work. We just raid them."

J.J. was a descendant of sharecroppers who came to Washington a few generations before he was born. Several million Black southerners moved north in the first Great Migration (1919–1940), and they landed in a stew of social problems. Hundreds of white scholars diagnosed the "Negro Problem" in book after book. Black families were poorly established on the landscape. They experienced a failure

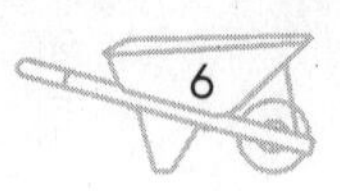

to root well. Black households did not thrive. Their communities suffered from blight. The problematized "Negro" in US history suffered chronically from root shock.

That's one story. But there are others.

Several million people moved north, and they brought seeds along with them. But they didn't bring all the seeds. They left behind the cotton, tobacco, and field corn, the plants that bound them to debt, drudgery, and destitution. The seeds they carried were of a garden variety—collard and mustard greens, beets, celery, tomatoes, peppers, and squash. They took cuttings from a mother's favorite peach and the hound dog ears of an uncle's pawpaw tree. Most every Southern migrant knew how to garden. They dug in and transformed the ground around them, growing vegetables and fruit while raising chickens for eggs and meat and hogs for pork. They produced a vegetable-powered wealth that got them through both world wars and the Great Depression.

The war ended. The "white noose" around urban Black neighborhoods squeezed in more bodies. Some people thought those (fertile, fecund) "chicken yards" were ugly and smelled too much of need. City leaders passed regulations banning gardens in favor of lawns. Encouraged by government-subsidized mortgages and low taxes, white people were drawn to suburbs. They swayed on the commuter train in one motion like stalks of corn in a field. Following them, one plant rooted: a delicate grass heavily irrigated and fertilized to stay green all summer. Due to its expensive fragility, turf grass signified affluence, order, and whiteness. After the Fair Housing Act of 1968, turf grass became the law of the land. It policed the boundaries of property values as effectively as any secret racial covenant. No federal funds, bulldozers, or police force could compete with turf grass for efficiency in marking out territory. Lawns spread to neighborhoods like J.J.'s, where memories of self-provisioning and community pig roasts faded.

I asked J.J. how he got interested in bees and fungi.

"I grew up around here." He lifted his chin to indicate a neighborhood in Southeast DC that, on writing, has one full-service grocery store for some 87,000 people and the city's highest rates of chronic disease.

"When the ice cream truck comes, everyone gathers round. Even if you don't have any money, you still go to the truck. There were about fifteen of us standing there, and a woman said that if anyone was looking for a job, she had one to apply for. I had my CV on my phone. I sent it off. I'd been idle about two months. Then I had a job."

"Now I teach kids from my own neighborhood." J.J. started to tear up. He stopped, words faltering.

Patricia Browne, director of the National Children's Center where the garden stood, comforted him: "That's OK. That's how it is. We got to say it how we feel it."

"Did you want to work with gardens and plants?"

"No, I didn't. I just wanted to get paid. My friends, they'd clown me. They'd say 'You're just growing flowers.' I'd tell them, 'No, I am growing broccoli.' They'd say, 'Same thing.' "

Later, I asked him why he cried.

"I don't know." He shrugged. "I don't know. Just made me think I'd come from there to here."

"What was there?"

"A black hole? Kids you knew getting in trouble and going to jail."

It had been seven years. J.J. was the director of Ward Eight community gardens in Southeast DC. Now J.J. had a pile of kids following him, looking up to him. He changed their lives like garden-variety plants changed his.

Tiny Gardens Everywhere offers a kaleidoscopic history. A kaleidoscope uses reflective surfaces tilted toward one another to show repeated patterns refracted through light. In this book, different shards of information, different times, places, and national histories go into the cylinder and rotate to reveal new patterns and connections. With each turn, the history moves from eighteenth-century England to fin-de-siècle Berlin, from Jim Crow Washington, DC, to World War II Germany, from postwar lawn enclosures in Memphis and Chicago to self-provisioning in Soviet Estonia, and, in the twenty-first century, to alternative food movements in Amsterdam and Mansfield, Ohio.

This book is part history, part memoir, and part manifesto. I look to the past to understand new horizons for cities and food systems in the future. Both cities and food systems will change radically in the coming decades given the precarity of today's industrial farming and distribution systems and the drumbeat of unbounded urbanization across the globe. In the past 150 years, urban dwellers got by and even thrived by creating the means for collective sustenance. Today, many things they invented are useful again. Turn the kaleidoscope and the story begins.

THE FIRST TURN

A Light in Dark England

Locked Out

THE SIGN SAID, "Welcome to the Northfield Allotments," but the gate was closed and bolted with a heavy steel padlock. We were in Ealing, a West London residential neighborhood. Peering through the grating of a six-foot fence, Marjoleine and I could see the first brave apple blossoms, red spears of rhubarb, and compost piles, some neat, most unruly. Between them, paths of bright green grass linked the large garden network.

Northfield, the oldest existing allotment in London, was founded in 1832 just after England suffered waves of machine breaking, arson, looting, and attacks on authorities and wealthy farmers across forty-five counties, the largest explosion of social unrest in English history. Trying to appease hungry, stone-throwing workers, the English elite carved up twenty acres of land on the outskirts of London into tiny provisioning gardens. Since its founding, the Northfield Allotments have shrunk to a few acres, land siphoned off over the years by developers.

That Sunday afternoon, there were only a handful of gardeners tending their plots. We finally got the attention of one. She came over but did not open the gate. We asked if we could come in for a look around. She said there was an Open Day in midsummer. We explained we would not be in London for that special day. We had come a long way just to see the gardens.

She said she'd love to help, but she could not allow entry to strangers. She was a new gardener and did not want to lose her plot by breaking the rules—she'd waited three years to get it.

We walked to another entrance, along a two-hundred-year-old hedge packed with blackthorn, hawthorn, apple, willow, and blackberry brambles. At another bolted gate, we stood looking in wistfully until a woman leaving the garden agreed to quickly show us around.

We walked through the gate to her plot. She had a few fruit trees in bloom plus some greens and herbs, peas coming up, and a drift of flowers in beds. But my eye was drawn to the rich, dark soils. Two centuries of gardeners' care showed in each springy step. I stopped myself from kneeling on the ground to touch and smell the earth, sensuous as it is.

"Don't play in the dirt!" Parents spend a lot of time training their children out of their habit of handling and sometimes tasting earth, though scientists are just starting to understand how contact with soil microbes supports healthy immune systems and childhood development. Maybe this instinctual allure is why waiting lists for allotments have doubled and tripled in the past decade. Maybe that's why, I thought with some bitterness, it's time to open the gates.

We continued our perambulation around the fence. A quiet, narrow footpath led along one edge. There, birds sang, mostly sparrows and doves, diving in and out of the hedge. I stopped and tasted a few leaves of wild garlic mustard. This "Class A noxious weed" was delicious, tender still in April.

We reached a commercial street and stopped at a café (where everyone spoke Polish) for a bowl of borsch. Nearby, in an old Anglican church, musicians were practicing for a Hindu ceremony. The public park (no fence, no gate) was jammed with children at play, parents, and young men, heads and knees together, nodding to a beat. The population in the park was starkly different from the few, all-white gardeners in the allotment.

Both public parks and private allotments served as compensation for a long process of enclosing public commons to make the urban and rural landscapes of today, in which territories are divided, fenced, and often guarded. Historians will tell you that once cities grew large in the nineteenth century, they broke the connection people had developed over millennia to their surrounding environments. City dwellers piped in fresh water and flushed dirt away. They purchased provisions, clothing, and materials from around the world. Urbanites skated along barely touching the earth on paved landscapes. For a long time, people associated cities with humans, and saw humans as somehow distinct from nature. But now scientists understand the human body not solely as a discrete vessel, shaped by ancestors' DNA, but rather as an ecosystem, a kind of Grand Central Station of microorganisms: bacteria, fungi, viruses, and invertebrates. Human bodies merge with the larger ecologies in which they live. Humans are in the thick of "nature."

The conventional historical narrative telling us that people flocked to cities to find jobs in new, large factories isn't entirely satisfying. Who would want a mind-numbing, physically brutal, and poorly paid job in a factory? When asked, nineteenth-century women said they would rather work long days in the fields than shorter days in a factory. Instead of being drawn to cities, rural people were pushed into cities from their ancestral homes. And they did not go willingly.

1830, Kent, England. A damp, twilight breeze snaked around the ivy-covered stone walls of an elegant manor house. From the stables, the neigh of a horse, and then, a scratching sound followed by a spark that illuminated the silhouette of a figure quickly stepping away. As flames spread from the barn, neighbors came running to help. Well, some helped. Others cheered from the darkness as the timber frame collapsed. A couple of people pulled out knives and slipped over to puncture leather hoses pumping water to put out the fire. A journalist for *Punch* claimed the blaze illuminated a "satisfied gleam in the eyes of a starved, sullen and revengeful peasantry."

The invention of coating the tip of a stick with sulfur first appeared in English markets in 1829. The "strike-anywhere" Lucifer match made fire portable, quick, and dependable. Good for house servants, cooks, and cigar smokers. Good for arsonists. Lucifer matches gave easier voice to grievances that had been smoldering in the breasts of rural people for decades.

When the fires started in 1830, Great Britain had just emerged from decades of conflict. English armies fought wars abroad for labor, markets, and territory, and they engaged in battles at home for the same. Beginning in 1750, Parliament passed Acts of Enclosure, making it much easier to privatize the common property of a parish without the consent of commoners. From 1750 to 1850, landowners used these acts of enclosure to appropriate six and a half million acres of communally managed property, a quarter of the country's cultivated territory.

By 1830, rural working people had reached a breaking point. Cut off from the pastures, forests, swamps, and gardens that had fed and sheltered them, they found it harder to feed themselves, to pay taxes and rent. Villagers had to work for wages that dropped each year as more and more people sought jobs. People lost their homes and shifted each year to new farms as hired laborers. Evicted commoners and poorly paid farmhands turned what weapons they had—arson, machine breaking, and looting—on the wealthy who were profiting handsomely from privatizing common land.

Historians have used the phenomenon of enclosure to debate the great changes to human and nonhuman life that occurred with the

Industrial Revolution, the scientific revolution, and the second agricultural revolution, all led by the English. The common view is that commons were and are a "tragedy" because, for lack of private property, people raced to get their share and overgrazed, overfished, drained aquifers, and depleted minerals. And that is certainly true. The biography of any robber baron shows how people take what they can get and keep it. But the Nobel prize–winning economist Elinor Ostrom, who was born during the Great Depression to a family of limited means, showed that when people are part of decision-making processes set up to govern shared resources with clear rules and boundaries, they can be adept stewards of commons. So, it's useful to take another look at how English commons worked before dropping them into the wastebin of history.

Common land use existed on every continent where people settled. In England, many legal categories of people had a share in the commons, but in short, the population could be divided into three classes. Peasants either rented or owned land that came with rights to use common pastures, wastelands, and forests. Wealthy tenants had large farms they rented from a landowner. Landowners claimed large portions of a parish, but by law they had to respect villagers' rights to use territory deemed as commons. Landowners were often absent for long periods. Until the eighteenth century, they rarely took an interest in farming beyond collecting rents.

Landscapes held in common were distinctive. Villagers called them "open," meaning the land was not divided by fences or hedges. People lived in cottages clustered in a village. A commoner could have a lease of sixty acres or just six. Commoners would pool together a team of horses and a plow. They had no individual control over what they planted but met regularly to devise a course of cultivation for all.

There are hundreds of books on English commons and their enclosure, many of which I read, but I still didn't have a good idea of what an open agricultural system looked like or how it worked. Finally,

I wrote to Sarah Tarlow, a historical archaeologist who in *The Archeology of Improvement in Britain, 1750–1850*, describes how it is possible to still see remnants of older agricultural traditions written on the landscape. I asked her if she would take me for a walk, and she kindly obliged.

On a rainy spring day in the English Midlands, we set off across a field planted for hay. On the other side of the hedge, the land was flat and marked by tractor treads, but where we stopped the pasture undulated up and down in long, gentle waves. In the furrows, water gathered, and we sank into mud. On the ridges, elevated by a meter, our feet were dry. Sarah explained that as pre-tractor farmers plowed, the moldboard pushed earth to one side, building up the ridge at the expense of the furrow. In the wet clay fields of the English Midlands, furrows served to drain ridges, making them dry enough to grow grain. When farmers got to the end of the row, they turned, kicking earth up onto a bulkhead. Villagers used the raised land of the "balks" to walk and lead their livestock. With no boundaries for private property, footpaths crossed the landscape wherever a person desired to go.

Standing in the field, Sarah showed me how farmers would sow or "broadcast" seed on the ridges, a hand in a bag, tossing grain left and right. Birds flocked around, taking their share of seed. Such haphazardly and closely planted crops could not be hoed, so other plants and grasses grew with the grain on the ridges. Herbs, wildflowers, and grasses also populated the wet furrows below and the higher balks. At harvest, reapers took what was in the mix, both cultivated and wild.

As the landowning class became interested in farming at the end of the eighteenth century, they viewed open-field agriculture with a critical eye. All those wild plants growing on the ridges they saw as a "nursery for weeds," while the furrows wasted space that could be used for cash crops. The English landowner and novice farmer Jethro Tull hated the idea of giving seeds away to birds. Tull fixed a box to a rotating wheel that dropped seeds in straight lines into furrows and

a harrow that covered the seeds with dirt. His new seed drill lined up plants in straight rows that could be weeded. He then devised a horse-drawn hoe to clear unwanted plants.

Commoners wanted no part of Tull and other elites' "improved" farming techniques. They used agricultural practices honed over centuries, a kind of traditional ecological knowledge, that varied from place to place. Tull, trained as a lawyer, was new to agriculture. He only turned to farming to pay off family debts. Villagers, eyeing the seed drill, did not understand what was wrong with sharing a few seeds with birds or hosting wild plants alongside cultivated ones. Sowing seeds closely prevented weeds from taking over crops. In the balks, on the edges of the fields where the birds flew and dropped seeds, wild cousins of wheat plants took root. The feral versions crossed with cultivated varieties and increased the genetic hardiness of the crop. Saving seeds for the new season, commoners choose hybrids that were best adapted to that particular spot and climate. This was a kind of spontaneous plant breeding before Gregor Mendel's famous experiments with peas.

The balks and furrows, left undisturbed, had other uses too. The wildflowers that grew there attracted pollinators. Bees and birds carried soil microbes that inoculated and enriched the terrain. Unplowed land on the balks offered undisturbed spaces for fungi to spread underground networks that fed nutrients to plants. Today, after decades of plowing to the very margins of fields, farmers are advised not to till at all and to ring cultivated crops with buffers of wild, flowering plants. Commoners' untidy open fields are returning to contemporary agricultural practice.

From the villagers' perspective, it made no sense to plant straight rows of one crop variety that then had to be hoed and was vulnerable to pests. The harvests proved them to be correct. By Tull's accounting, his seed-drilled fields produced only about half the yield per acre of commoners' traditional broadcast sowing, but, if the farmer had a lot of land, as Tull did, the seed drill and horse-drawn hoe produced greater profits because he saved on seeds and labor.

Commoners, usually women, also kept kitchen gardens near their cottages, where they grew plants for food, medicine, textiles, flavored ales, and tea. Tansy and wormwood stopped the bugs biting. Roses, lavender, and rosemary scented the soap. Botanical knowledge was passed on in verse, which is easier to remember, so a gardener knew to place raspberries next to roses, that September was the best time to transplant gooseberries and currants, and that sowing edible weeds like fat hen and thistle between the crops serves as a living mulch. All plants, including some weeds, went into the pot for the daily soup.

A quarter-acre plot would hold several hundred species of plants. Plants work together to great effect. When a threat is near, they send off alarms to warn others. They release chemicals that kill insects, or they issue aerosol compounds that attract pests' natural enemies. The more kinds of plants, the more defenses to fend off the blights and insect infestations that ran unobstructed through landowners' large mono-crop fields.

In a parish, commoners had historic rights to use vast tracts of land called "wastes," a misnomer because the wastes were rich in resources. Commoners relied on swamps, fens, forests, and heaths for fuel, gravel, stone, and wood to make tools and to build and repair houses. Women and children especially turned to wastes to enrich the family diet with small game, birds, fruit, mushrooms, and nuts. From fens, they gathered fish, eel, and waterfowl, and collected berries and greens from hedgerows. The list of foraged materials goes on and on. Commoners were considered poor, but had a richly varied diet at a time when the English elite had come to favor meat, wheat, and sugar, the precursor to today's fast-food diet. If one crop failed, commoners had others to harvest. A mixed economy on a patchwork of land served as an insurance policy against hunger.

Commoners worked for wages, too, usually as farm laborers.

Women joined for the harvest to reap and thresh. As teens, boys apprenticed on farms or with artisans, and girls as dairy maids. Peasants usually married after their apprenticeships in their mid or late twenties. Delaying marriage kept family size in check.

The idle rich disparaged commoners' small-scale farming and foraging as "the trifling fruits of overstocked and ill-kempt lands." How wrong they were. A woman's kitchen garden, foraging, and family cow, pastured on common meadows, could earn as much as a male "breadwinner's" annual wages, or the same profit as a wealthy farmer with eighty acres after he paid for horses, fertilizer, and labor.

As Sarah and I continued our walk, tall, thick hedges flanked every field. Hover over the English Midlands and you see emerald green pastures embraced by such dark green hedges. There are lots of reasons to cultivate a hedge. Hedges shelter plants and animals from the wind, protect against the neighbors' animals, and manage social relations. Most hedges today are populated with hawthorn and blackthorn, thorny, quick-set hedges, that laborers rolled out like carpet in the nineteenth century to mark borders of newly enclosed fields. But not all hedges are created equal. Older hedges are recognizable for their far greater range of plants.

"The ancient hedges were a whole pantry," Sarah exclaimed.

Running through villages and along roads, pre-enclosure hedges made up lateral food forests, planted and maintained for easy pickings: crabapple, cherry, hazelnut, rowan, elder, hundreds of varieties of roses, wild blackberries, ferns, hedge parsley, pennycress, bath asparagus, gorse, medlars, and medicinal plants whose names describe the remedy—goutweed, kidney vetch, bladderseed, and lungwort. Villagers knew how to cut back trees in a hedge so that each year the coppiced stems sprouted many more branches, which they then trimmed for tools and fuel. Many people know the history of deforestation, but far fewer know about the management of forests and groves as renewable and possibly immortal resources.

Landowners seeking cheap labor thought carefully about how to deprive commoners of an independent means of subsistence. Better, they advised, not to build hedges from medlars, a thorny shrub with a fruit picked in winter that tastes of cinnamon and apple. Medlars were unwanted because, as Minister Thomas Rudge said in the late eighteenth century, "the poor are already too prone to depredation and would still be less inclined to work if every hedge furnished the means of support."

If economic rationality alone served as the driving reason for enclosure, then privatizing wastes, land considered no good for farming, would have made no sense. But these "wastes" gave a poor household so much for free. With gifts from common wastes, one elite observer noted, "They get the desire to live . . . without labour, or at least with as little as possible."

How did commoners devise a landscape that produced food, fuel, and shelter with a minimum of labor?

They did not work alone. Commoners drew on a whole network of alliances.

The Dung Community

OPEN-FIELD FARMERS had a relationship with their soils that steered close to veneration. For English peasants, the earth was alive, the soil to be nursed and humored. Peasants broadcast offerings to Mother Earth—the blood of cattle, wood ashes, and salt. Before plowing, they rubbed fennel on the plow's cutting edge and asked the earth's forgiveness for disturbing it. Working to nurture the life they perceived in soils, they cared a lot about the dung heap. It stood in the center of the farmyard. A robust, steaming pile of manure symbolized a healthy farm.

Elite critics often disparaged the scrawny look of commoners' sheep. Most assumed that villagers did not know how to breed their sheep properly, but commoners selected sheep not for the quality of wool or meat, but for how well they pooped and when. For them, the animals' greatest value was in restoring fertility to the fields after a harvest. From April to September, peasants fed sheep by day on the down and "folded" or penned them by night on fallow fields. They moved the pens nightly, covering about half an acre to eventually manure the whole lot. In breeding, open-field farmers regarded a sheep's "folding quality" (the propensity to drop manure after being penned at night) as more important than how much wool, meat, and milk they gave. If you cared for your soil and about saving labor, breeding for manure made a lot of sense. Manure draws a world of life to itself.

A child leads the cow to the common pasture to graze. The girl sits down and watches puffy clouds float by. The cow gets to work, pulling grass, aster, clover, thistles, dandelions, and chickweed, and starts chewing. The cow ruminates, spinning out a green cud that when gulped for the last time churns through four stomachs that are specialized in the art of breaking down plant fiber. While ripping up plants, the cow swallows several cups of soil a day, and with it, millions of soil microbes. These tiny creatures make their way into the cow's gut to join a mass of other parasites—gut bacteria, viruses, fungi, and worms.

Tossing her head, the cow lifts her tail. Before the dung reaches the ground, a fly has already planted her eggs in it. Beetles dive into the pile, blindly tunneling in search of bacteria and fungi to eat. They scatter the flies' eggs. More insects and birds swarm around the rich plate of food that is fresh dung. Birds peck out seeds and sail off to deposit them elsewhere. Wasps and spiders feed on fly larvae. Mites arrive. From below, earthworms set to feasting. As the pile breaks down, fungi finish the job of turning the dung into a rich, earthy humus. The fungi release spores that float upward with air currents and attach to water molecules that form clouds that the child shepherd watches as she daydreams.

The cow and the dung community work together as an efficient biodigester, each creature eating the refuse of others, to break down stubble and plant matter on fallow fields, speeding the regeneration of soils for the next planting. Commoners had a rule to mow no earlier than October. If they cut hay too soon, too vigorously, they destroyed insect eggs laid in plant stalks, crushing them. They grasped how useful the dung community was. In places such as Australia that have no native dung community, cow manure doesn't break down and it piles up disastrously.

The English elite approached soils in a curiously different way from commoners. Rather than bending over, picking up a handful of earth, rubbing it between two fingers, and smelling and observing it as a commoner would, they approached soil as "dead matter" to be fixed by modern chemistry. Leafing through nineteenth-century manuals for "improved agriculture," I found that farmers were advised to purchase chemical additives called "mineral salts" to balance soils. The idea was that to get the soil chemistry right, no particular local knowledge was needed. All a farmer had to do was dump the right mix of purchased fertilizers, and crops could reliably grow anywhere in an expanding empire—in rainy England, in tropical India, on the high dry American plains, or on the Argentine pampas.

With the shift of land into the hands of wealthy farmers, soil depletion became a real problem in nineteenth-century England and other places where farmers engaged in "improved" agriculture. Plowing up former fertile commons initially increased harvest yields. But pests and blights rampaged through rows planted with one crop and extensively weeded. Heavier plows plunged deep into the earth, killing off microbial life. As soil organisms died, fields dried and turned gray. When yields fell, wealthy farmers purchased more fertilizers to add nutrients to plants. They adopted Tull's seed drill, and plowed more frequently and more deeply, intensifying erosion. As topsoil rolled away, wealthy farmers laid down more and more imported fertilizer, often guano mined in South America. Struggling against soil depletion, the volume of fertilizer in England multiplied from an average of seven pounds per acre in 1820 to ninety-six pounds in 1870, but with no gain in yields. Soils depleted of microbial life fall apart. They hold water and fertilizers poorly. The more farmers added fertilizer, the more the

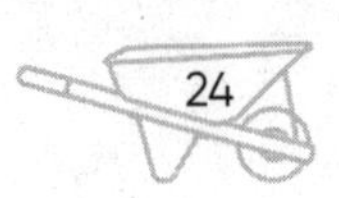

added nutrients leached into water tables, bypassing crops. With yields leveling off while the costs of agricultural inputs increased, landowners sought to enclose more commons and to free themselves of the burden of villagers' common rights. That, they said, would be progress.

The seventeenth-century philosopher John Locke first came up with the idea that those who improved "unassisted nature" increased productivity and generated new value. "Improvement" came in the form of draining wetlands and consolidating smallholdings into large fields. With these ideas in vogue, the ruling class divided the populace into two categories. There were enlightened, dynamic people, like themselves, who embraced "progress," and commoners who sat tight and refused to budge out of ignorance or slothfulness. These were "Luddites" who destroyed new machines they did not understand and violently opposed any form of improvement.

With these ideas, elites came to see commoners and their mixed economies as representing an earlier, "primitive," stage of human history, as if the clock for them alone had slowed. Why would an advanced nation like England tolerate such backward people? It made as much sense, a mid-eighteenth-century observer noted, to preserve commons and open-field agriculture as it did to leave North America to the Indians. Bigger farms and cheap labor would lead to greater efficiencies, higher yields, cost savings, and greater profits. Revenue from agriculture would help other businesses at home and abroad in the Empire.

Commoners, they said, did not take their children to school, missed church on Sundays, and, worst of all, did not toil with proper zeal. Some only worked four days a week or stopped midafternoon. A landowner complained: "If you offer them work, they will tell you they must go to look up their sheep or take their horse to be shod that he may carry them to a horse-race or cricket-match." A major advocate for enclosure, Arthur Young called village women "lazy" and improbably charged, "they do nothing but bring children and eat cake." Joseph

Townsend reasoned that for the poor "only hunger can spur and goad them on to labor." A Mr. Bishton from Shropshire dreamed in 1794 about the hour when the commons were enclosed and "the labourers will work every day in the year, and their children will be put out to labour early." For the enclosing class, commandeering common land solved a labor problem and a moral one at once. The Reverend Thomas Malthus, who dabbled in population statistics, insisted that self-subsistence ruined commoners, while the market educated. The discipline of wage labor, he promised, would purge the poor of their sin and sloth.

I am a commoner myself, in a fashion. I have gardened all my life, but until recently I didn't have any land of my own, so I got on the waiting lists of community gardens. I often rubbed against the "community" part of these urban commons. All those people, rules, negotiations, and obligations. There would be disputes over whether dogs were welcome or not, or what to do about unhoused people who swiped tomatoes. In Washington, DC, my friend and I got kicked out of a community garden in a posh neighborhood for having too many "weeds" (we planted and cooked up that "weed"—self-seeding and nutritious shiso). After that, Ryoko and I joined up with a few neighbors to build a new community garden on a patch of unused land alongside an elementary school across the street from where I lived. We didn't have much room and only half a day of sun, but we did what we could. We suspected lead in the soil, so with a group of neighbors, we constructed a dozen raised beds that, though the standard size, were always too confining.

Ryoko, a talented horticulturalist, started planting outside the beds, along the sunnier edge of the fence, where she had built up the soil by adding compost, leaves, and garden scraps, elevating the plants from contaminated soil as in a raised bed. On the garden online discussion list, a member asked who was doing that and why. I sensed trouble, but Ryoko replied that she was planting edible and beautiful plants for everyone to harvest and enjoy.

Suddenly, with this invitation to create a common growing space, everything changed. I too poked some seeds along the fence and added mulch and compost. A couple dug up some ground and cultivated hops for home brewing. The hops grew madly, leaning against raspberry bushes that another gardener had planted. A dull, orderly space became wildly interesting once Ryoko unleashed our plants from the little raised beds that had been holding them back. Each morning I was propelled outside to see what had happened overnight.

One morning, I stopped to take a look at some harlequin beetles gorging on kale. Pitted and brown, the kale took a drubbing each summer as the weather grew hot. But next to the kale, drought-resistant sweet potatoes thrived. I continued past a stand of beans that an eighty-year-old gardener had given me. When he handed them over, he had murmured something about Jack and the Beanstalk. I realized he wasn't joking. Those beans climbed up the ten-foot school fence and out of the yard like truants, one leg over the wall. I headed into the woods of Rock Creek Park to take my dog, Jupiter, for a walk. As I descended through the cool air toward the creek, my thoughts turned to garden plans.

The community garden was growing fine, but I still had many other plants I wanted to put in the ground. The District of Columbia had recently established an Urban Farming Land Lease Program that made public lands available to residents for cultivation. Perhaps I could get land through that program?

I called up the director of the Office of Urban Agriculture, who told me that Washington was one of only three US cities with such an office, but that, other than her salary, her bureau was unfunded. She explained that there was so much red tape involved in getting just a portion of an acre for urban farming that only two parties had managed to cut through the bureaucracy in four years. That was disappointing. I knew my gardening hobby was not ardent enough to navigate through the labyrinth of city hall.

But next to our row house was a slope of land wedged between an asphalt parking lot and the sidewalk. The little plot attracted dog poop, weeds, and fast-food wrappers. I put on gloves and cleared out the refuse and weeds, leaving a few stands of perennial mountain mint, garlic chives, and tulip bulbs planted ages ago. I stepped back and looked at my work. The slope got a lot of direct sunlight. It was dry. Something Mediterranean might prosper here. Tomatoes and peppers? A fig tree?

I bought a fig sapling, took some compost from my bin and worked it in. While I was at it, I sowed tomatoes, peppers, arugula, chard, and spinach. Who knew? Something might happen.

It didn't take long. The fig grew inches a day in the hot sun next to the baking asphalt. Figs are an amazing urban tree. They like tight crevices and can tolerate drought and air pollution. The tomatoes and peppers kept up too. In the shade of the fig, the greens thrived.

One day I was on the slope, trying to figure out how I could keep the soil from washing down onto the sidewalk. A tall young man with a strawberry-blond ponytail walked up and introduced himself as Wesley. He had recently moved into a group house next door.

"I wouldn't mind a little piece of the action here," he said, nodding at the garden.

Sure thing. I was happy to have company.

Wesley proposed we harvest bamboo growing nearby to weave terraces to hold the soil. We got to work cutting and braiding stalks of black bamboo into fences. We thought it looked good. Wesley's day job is as a gardener, and he had plants left over from a client. We popped in sorrel and strawberries. Among them, tulips, purple baptista, and gaudy yellow daffodils bloomed like splashes on a painter's palette.

But the little garden didn't sit right with a few of the neighbors who lived in the apartment building behind it. Some had dogs and would come out in the morning, cross the parking lot, and let their canine charges use the sidewalk garden as a toilet. Dogs, like humans, are omnivores, and their feces can contain viruses, harmful bacteria, worms, pesticides, and pharmaceuticals. To use dog poop safely as fertilizer, the dog would need to have a pretty clean diet, and the poop would need first to be broken down into compost.

We never really solved the problem of the slope as either garden or toilet. And it stewed.

A few years passed. The tree grew and produced baskets of sweet purple-brown figs in the fall. The tomatoes, peppers, greens, and herbs saved Wesley and me trips to the grocery store. The two of us made friends over that garden, though we differed in age by thirty years. We liked spending time together and so we gardened a lot, which meant the little plot looked great. People would walk by and name the plants or just say they looked pretty. Sitting on my front porch, I'd hear them. So would Wesley, sitting on his front porch on the other side of the wedge garden. We'd raise a hand to each other, proud of the tiny domain we had created.

Not everyone in England thought enclosing communal land was a good idea. In the late eighteenth century, Thomas Spence, a working-class writer and bookseller, witnessed how the wealthy appropriated common property so that "no one could claim a right to so much as a blade of grass, a nut, or an acorn without the permission of the pretended proprietor." In a speech in 1775 to the Newcastle Philosophical Society, Spence laid out his alternative order: "the country of any people in a native state is properly their common." No one had the right to sell common lands and other resources upon which everyone depended. "For a right to deprive anything of the means of living" is "a right to deprive it of life."

Spence declared that inhabitants should come together, form a corporation, and declare the land where they lived to be theirs. They would pay rent to the parish, and that revenue would go to local public expenditures, including poor relief. Spence was no proto-Bolshevik with a desire to tear property from the wealthy. He reasoned that everyone would eventually want to join their neighbors in a cooperative venture. In the meantime, he dreamed that wastes and commonly held parish land could be rented out in very small farms for "victualing" (self-provisioning), and that way a great many men, women, and children would find employment and sustenance.

After his speech, Spence was expelled from the Newcastle Philosophical Society. He spent time in jail for his views and died as poor as he was when born. But others also began to question the new order. With a raised eyebrow, Adam Smith remarked in *The Wealth of Nations* on the exceptional rents commanded by enclosed land. Prime Minister William Pitt asked in 1809 about the wisdom of enclosing an acre of wheat, which fed a family for a year, to pasture a cow for just a few weeks.

By the nineteenth century, the social impact of the new round of enclosures was clearly visible. Peasants who once had subsisted from the commons and wastes had to make do with low-wage jobs that dried up in winter. With no common pastures and forests, they had to give up the cow and pigs. Prices rose as formerly self-provisioning people bought staples in shops. Families racked up debt for rent and groceries. Competition for work was stiff, so employers lowered pay. With fewer apprenticeships and jobs that had delayed the age of marriage, teens married a decade younger than their parents' generation. As the years of married fertility doubled, so did the number of children. Parents hired children out for jobs from age eight to twelve, their tiny wages contributing to the family's diet of potatoes with a bit of fat. By 1800, "surplus" populations stood, heads down, tags pinned to their chests at hiring fairs.

New vagrancy laws made it a crime to "refuse gainful employment." Other everyday actions also became criminalized. Picking up sticks in the forest, taking a walk in the woods, making a shortcut across the field, gathering greens or mushrooms, or trapping a rabbit in winter when the family had nothing to eat all became criminal acts.

The confidence of the terms—*progress*, *improved agriculture*—expresses the hubris of a class that prospered from the appropriation of land that enclosure enabled. The long history of enclosure brought into the English vocabulary two new terms—*pauper*, which describes a person with no home and no means to make a living, and *private property*. As the wealthy accumulated property, English peasants went from foraging, grazing, and tilling to getting tickets for food or being checked into prisonlike workhouses. The closure of the commons and the flood of working poor into cities revealed that the elite had long failed to pay laborers a living wage. Rural people had made up the difference with kitchen gardens, by using common pastures, and from gleaning the edible landscapes they maintained. "Vigorous accumulation," geographer Jason Moore writes, "depends on the existence of people and other creatures whose cost of sustaining themselves are kept 'off the books.'"

Agricultural reformer Arthur Young had once advised pressing the poor into the army. It would be doing them a favor, he had reasoned, for them to be killed by cannons rather than to be forced to be idle. In 1800, he suddenly changed course. He saw the hungry families and noted the rise in rent and staples. "To not speak up in their defense," Young wrote, "would have been a wound to any man who had viewed the scene which I have visited."

Young's solution was simple, but also a cruel reversal of the decades he spent lobbying to destroy the commons. He noticed among enclosed parishes great tracts of former waste and commons "of which little or no use is made." It struck Young that rather than providing welfare for dispossessed families, it would make more sense to hand over wasteland so villagers could support themselves. Young calculated that just three-quarters of an acre for each family would suffice.

This proposal was outrageous. Young wanted to give a portion of the land that had just been wrested from commoners back to them. Members of Parliament quickly iced him out for this show of sentimentality. Instead, a Parliamentary committee advised solving the problem of poverty and "excess populations" with mass emigration abroad.

But the London surgeon Benjamin Wills backed Young's idea. People don't want to be forced by hunger to leave their country, Wills insisted. Instead of colonization abroad, he proposed "home colonization." Wills founded the Labourer's Friend Society and wrote a general enclosure bill that would allocate a certain proportion of enclosed land "for the nation." In his plan, commissioners would divide a hundred acres of former commons into small parcels he called "allotments," which would be let out rent-free to poor people for thirty years.

Malthusian leaders of the day dismissed Wills as a harebrained radical. But he had a band of traveling agents promoting allotments as a solution to tax increases to support the poor. In the 1820s, several score of wealthy philanthropists and clergymen divided land they controlled into small plots for former commoners to till for subsistence. They became known as "allotment dukes." These experiments showed signs of success. Hundreds of families cultivated quarter-acre plots. People who would have starved instead had blooming kitchen gardens.

Still, most wealthy tenant farmers and landowners hated the idea of allotments. Editors for the *Economist* called them "a dangerous scheme." Farmers and landowners feared that they stood to lose their bargaining position with farmhands if people had the luxury of feeding themselves. As the report of the Poor Law Commission put it, "The more they work for themselves, the less they work for us."

You can see in this equation the narrowing of the definition of "work" to mean only wage labor, gradually gendered male. Other labor that men, women, and children did "for themselves" lost value.

With both landlords and wealthy farmers against them, allotments looked like a dead letter.

One day, our neighbor Constance decided that Wesley and I had no right to the garden on the slope. She said it belonged to the people in the apartment building. It was, after all, in front of their building, where they paid rent. They were going to cultivate it.

"Sure, the more the merrier," Wesley said, but Constance was not talking about joining our garden project; she wanted to take over. We were getting evicted.

Constance pointed out that Wesley and I had other land to plant.

That was true. We both had raised beds in the community garden across the street and we'd started putting in strawberries, berry bushes, figs, persimmons, and pawpaws in the green spaces around the school. Wesley and his group house members had just managed to buy their rowhouse. "As landowners," she said, "you think you have more rights than us tenants."

Wesley and I had taken wasteland and made it into a garden that produced fresh food. We felt our labor and care made for a kind of ownership, regardless of who held the title. Constance didn't want a history lesson, but I offered one anyway. In Roman law there were three forms of property rights. The first form, *usufructus*, gave people the right to use property and enjoy its fruits. The second form, *abusus*, gave the owner the right to destroy property. (I didn't say, though I thought it, that her dog was making use of that right.) Only the third form of land use, *dominium*, involved the right to exclude all others from the property. And clearly, none of us had that right.

But Constance wasn't interested in my pedantic explications. What she wanted was simple and reasonable. She lived there, and she had concluded that she and other tenants had the exclusive right to use the wedge space as they saw fit. After all, they paid rent that kept the building and our disputed plot from the tax collectors. Constance had no use for our suggestion that it was a commons that we could enjoy and manage together.

The right to own property and exclude others stands foremost in the American understanding of land tenure. US landscapes are alive with boundary markers, fences, walls, locked gates, and signs warning, "Private Property, No Trespassing." Yet for most of human history, exclusive ownership of property would have been a strange conception.

But what about the soils and plants? Who owns them? Wesley and I cultivated and cared for them, but once Constance made her claim, we lost our ability to look after them. Noticing that the strawberries were just about to ripen, I felt keenly how property is a form of power, with boundary lines serving as the vehicle to exclusion.

We spent a sad day unearthing and moving a few plants. Wesley and I also shoveled out some of the soil we had built from composted kitchen scraps and mulch. I went around feeling hollow for several days, mourning the plants I had set in the ground with bright, spring expectations.

Allotted Survival

AFTER THE FIRES, machine breaking, and uprisings, the proposal to grant the working poor small gardens to sustain themselves suddenly made a lot more sense. In 1831, Parliament passed three acts to enclose up to fifty acres of waste in a parish to rent back to the poor in small plots of usually a tenth of an acre. At first, working people did not trust the allotment scheme. They harbored suspicions that the landlords had no interest in helping them. The land they were offered was usually in poor condition, needing to be drained, dug, and fertilized. Allotments also had to be fenced or hedged, the same expensive infrastructure of enclosure that shut cottagers from their common lands. Commoners understood that allotments were not an innocent architecture.

But those who risked it found they could grow enough to save about 20 percent of their household expenses. Their testimony convinced others. Waiting lists for plots grew. The landowner Stephen Demain-

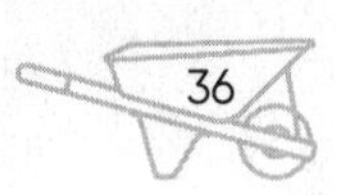

bray had so many requests, he divided eight acres into ninety-five portions. Only in places where wages were high was demand for garden plots low. By 1845, of 614,000 acres of enclosed commons, 2,223 acres were "allotted" to former commoners for self-provisioning.

Hold on—less than 1 percent of hundreds of thousands of acres of commons was given over to displaced communities in the form of allotments? That is poor compensation for the complex of field, forest, and pasture that commoners lost.

Of course, the purpose of allotments was not reparation. The elite had no intention of returning commons to commoners. The goal was merely to keep laborers from starving and rioting, while lowering taxes (the poor rate) to support the jobless workers. The philosopher John Stuart Mill cynically quipped that allotments were a strategy to have people "grow their own poor rates." To ensure that people would still work when needed, landowners carved fields into tiny gardens "insufficient for a full subsistence."

And more was lost than land. While commoners had elected their own leaders and managed communal territory collectively, landlords ruled allotments. To get a plot, field wardens took into account family history, criminal record, employment, church attendance, and the number of children. No one considered "saucy" or "lazy" got a plot. Nor did men who joined labor unions. Men received allotments; only rarely did women get them. Women, who had long cultivated kitchen gardens, were edged out as male heads of households came to dominate the family growing space.

Landlords wrote up a long list of rules designed to instill discipline and dependency. No gardening on Sundays. No more than half the land in wheat (a cash crop). Plots were to be kept neat and clean. Allotment holders could not sell their produce, nor work so much in the garden that they neglected wage labor. To keep men out of the alehouse, landlords might locate the allotments a two-mile walk away. People convicted of committing crimes lost their plots. That rule was effective. "The men were more afraid of losing their allotments than going to jail."

Tiny allotment gardens, as they emerged in this harness of regulation and spatial segregation, became a scale model of private property, possessive individualism, social hierarchy, and paternalism, ideologies that ruled the day. As two of the first allotment historians put it, "Compared to common rights, allotments were a symbol of dispossession."

Coming to understand this history, I grasped that the heavy locks on the gates of the Northfield allotments in East London and the other lush allotment gardens circling London were rooted in the slow, thoughtless malnourishment of the poor, a descent that was disguised in tales the landed class told themselves of "natural laws," improvement, progress, and charity.

But surely the misery in the decades of consolidating small plots and common territory into large farms was worth it. Improved agriculture fed new, large cities and that wealth powered the Industrial Revolution, which has made our lives so comfortable. Right?

In the early twentieth century, the exiled Russian prince, ecologist, and geologist Peter Kropotkin took a walk in South Devon. He was struck by abandoned fields, orchards, and cottages going to ruin. He saw field after field with nothing but grass and thistles. The soil was gray and dry, the topsoil worn down so clay subsoils poked through like holes in a moth-eaten sweater. By that time, England no longer fed itself. Improved agriculture had not worked out.

As farmers abandoned fields, English merchants turned to other parts of the globe for cheap food. In Britain, 80 percent of wheat and 40 percent of meat came from abroad. The new imperialism of the nineteenth century drained much of the world's most accessible fertile land. The rate of enclosing and plowing under what was called "virgin" territory in India, Asia, and South and North America outpaced even the rate of population growth, which was unprecedented. The farmers did not go easy on the land. They cut down forests and ripped through grasslands, sent topsoil sliding downhill and over riverbanks to create new geological features such as the artificial Little Grand Canyon in Georgia and the elevated Mississippi River. Within a few years of homesteading, farmers in Kansas noticed their soil thinning. By that time, dust storms and famines caused by dependence on mono-crops, heavy plowing, and soil erosion convulsed Russia, China, Austria, Sweden, Ireland, and India.

Nineteenth-century thinkers obsessed over the problem of degraded soils. Charles Darwin in his dotage became fascinated with the soil-building properties of earthworms. At the end of his life, Karl Marx determined that all sources of wealth grew from the soil, first, and workers only second. He read about soil science and fretted over the "metabolic rift," which he characterized as farmers' failure to return nutrients to the ground they worked. Rosa Luxemburg spoke of the theft of earth right out from under workers' feet. Early twentieth-century conservationists wrote often about "desertification." Nothing short of a miracle, they worried, would halt the tip toward soil exhaustion that threatened to turn all of planet earth into a vast desert.

Paradoxically, as farmers "raped virgin soils," "spade" gardeners on allotments thrived. Many people noticed that "the poor man's crop never failed." Even when allotment gardeners took over exhausted land that the farmer no longer found profitable, they produced astonishing returns. Allotment historian Jeremy Burchardt calculated that small gardens delivered double the yield obtained by large row-crop farmers.

A lack of resources propelled gardeners' exceptional fertility. A spade cost half a crown, about a day's salary. With that investment, a family got digging. Allotment holders understood that the gardener who plows "is rarely a successful or permanent cultivator." Although improved agricultural manuals recommended plowing often and intensely to dredge up minerals and eradicate weeds, folk wisdom held that spade farming led to higher yields. But how could a wooden pole fixed with a slice of metal outperform the plow and later the first tractors?

Plows and horses are heavy, tractors more so. A tractor rolling overhead is like a mine collapse. Little fungal and microbial colliers digging below for nutrients get starved of air and water. Repeated plowing churns through the life of soils, cutting fungal connectors and exposing microbial life to the elements. These subterrestrial beings can handle exposure to light and air once or twice, but eventually, fungi and microbes die off with persistent undressing from plowing. As they expire, microbes release nitrogen, which plants feed on, but most of the nitrogen unearthed from plowing is wasted as it melts into the atmosphere.

But with simple shovels allotments flourished. By 1873, one in every three male agricultural workers in England had an allotment. Urban workers also acquired them. In 1880, the National Agricultural Labourers' Union included allotment gardens in their platform, along with the right to vote and collective bargaining. Unions embraced allotments because they changed the equation in labor disputes. When miners or factory workers went on strike, families could fall back on garden provisions to tide them over. Workers' gardens cultivated organizational skills and new political tools so that urban and rural protestors could engage in open politics and trade unionism. In 1887, an allotment candidate won a seat in Parliament, and later that year Parliament passed another allotment act that gradually established gardens on public land, which meant lower rents and more permanence than allotments on private land. From 1860 to after World War I, more applicants desired allotments than could get them. The number of allotments grew in twentieth-century England to reach a half million plots.

Tiny gardens, as meager as they were, built alternative political dominions. The spade cultivated care. Care extended outward from plants and soil to more inclusive politics. One definition of freedom is being able to re-create oneself through a connection to others. After the commons were wiped away, allotment gardens restored in some measure the ability for working-class people to make their own worlds.

Thistles and chickweed were overtaking a few struggling strawberries and flowers on the wedge garden on the slope. I ran into Constance. She agreed the garden she had commandeered from me and Wesley did not look so good.

We had the same problem in our community garden. Many gardeners showed up to put some store-bought plants in the ground and then left. One gardener in our community was a well-known sleep doctor, famous for advising parents to let children cry it out in their beds. He, unfortunately, adopted the same approach with his plants, and his plot was quickly overrun with weeds. Wesley and Ryoko, talented gardeners, ended up looking after the doctor's plants and those of others.

Curious if such neglect was a trend, I made a tour of community gardens in DC. I saw a lot of raised beds parked on old tennis courts and along bike trails. Many beds looked abandoned. Finding people to build gardens is easy, but securing volunteers with the patience to weed and water and beat back rats and aphids is another matter. A gardener who shows up frequently can stop an aphid explosion by plucking infested branches or blasting them with soapy water, but if gardeners drop off plants and forget to check on them, the sad results of wilted kale and broccoli can be discouraging.

But some gardens I visited were astonishingly beautiful. In Southeast DC, where fresh produce is hard to come by, Rosie Williams, a teacher's assistant at the National Children's Center, managed a gorgeous community garden. Williams brought in seeds she had saved from her native Trinidad. With the children and volunteers, she planted greens next to onions, basil alongside tomatoes, and long rows of broccoli and peppers. She cultivated flowers such as yarrow and Queen Anne's lace that attract insects that eat aphids. Rosie got a beehive going. She pulled bushels of produce out of the garden and handed it to the cook in the school for disadvantaged and handicapped children. The school offered produce for sale at a Saturday farmers market. Some neighbors didn't know what to do with chard and leeks, so Nelson, the school's chef, set up cooking demonstrations. By the end of the second year, Williams and other volunteers had grown 2,000 pounds of food on a quarter acre.

When Kate Tully moved into a row house in Northwest DC, she noticed a community garden on her block. The garden had some sorry-looking raised beds. Tully wanted to grow vegetables but didn't know who was in charge. She scouted, waiting for some sign of life, and finally saw a man unlock the gate and enter. She ran out to meet him. Steve Coleman, director of the nonprofit Washington Parks and People, wasted no time recruiting Tully to become the garden coordinator. Tully, who had just started a new job, reluctantly agreed.

She could see from the gray, weedy beds that she wanted nothing to do with little private plots. Tully grasped that people garden for all sorts of reasons, but one of the chief motives is to be part of a community. Tully redesignated the raised beds as communal. She and her husband planted berry bushes and organized a composting station. They set up a group for weekly trash walks around the green.

Tully told me, "At first, there were so many bottles and needles that we got a special container to dispose of them. But as the garden got better, we found less and less trash."

People respected the space. Tully led work parties every Wednesday evening and Saturday. People came. They were happy to spend time together. Some had energy but little knowledge. Others had gardening know-how but less mobility. The gardeners divided into groups, one working the eggplant bed, another tending to milkweed for butterflies, a third team turning over the compost pile. Some neighbors just liked to hang out around the firepit, enjoying the shade and flowers.

When the Covid-19 pandemic hit, more volunteers piled in. Tully's mailing list grew to include 340 people. She worried that the common harvests would be too small for the scores of volunteers who showed up weekly. She would weigh the harvest and spread it out on a picnic table.

"I drew up charts with hours worked to determine how many pounds of food for each person. It was so complicated. I really worried about it, but I stopped after a few weeks. No one took too much. Everyone was careful to make sure everyone else had enough. And there was always food left over."

The Columbia Heights Green has just short of a tenth of an acre of garden beds. That is really tiny. In 2021, Tully and her group grew 2,600 pounds of food, sharing it among volunteers and giving the rest to two soup kitchens. Fresh food, no packaging, no chemicals, no shipping, and neighbors who know each other. What could be simpler?

I looked around in the historical record for other cases of people coming together to form collective gardens. That quest brought me to Berlin.

THE SECOND TURN

The Berlin Commune

4 The Green Ghetto

NINETEENTH-CENTURY GUIDEBOOKS warned travelers to steer clear of Berlin's working-class neighborhoods. A visitor could be robbed in the tenement slums. But one day in June 1872, journalist Max Ring set out on foot with a friend to find a shantytown known as Barackia. The two gentlemen strolled along the Landwehr Canal, past fine new houses on streets that developers had recently plotted out in a crazed real estate boom. They saw the new gasworks that lit up Berlin's better addresses. Boatmen poled barges piled with goods along the canal, which gave off a powerful stench on a warm summer day, the waterway brimming with human excrement that flowed from the city's cesspools and gutters. As they walked, streets gave way to a meadow and sandy potato field.

Just where urban life appeared to have faded away, Ring and his friend came across a neighborhood of tiny streets, dirt paths really, peppered with crude huts slapped together with planks covered with whatever the builder could lay hands on—tar paper, reeds, cardboard, felt. Each hut had a garden. Phlox, carnations, peonies, and asters shone in the sunshine. Bush beans, leeks, and celery grew in beds. Chickens pecked in the dirt. Penned rabbits and pigs chewed on scraps. Dogs slept in the sunshine, their backs to steaming piles of compost. The scent of frying onions wafted along the footpaths. The stench that had stalked Ring on the canal dissolved into a flower-scented serenity, an "idyll," he called it.

The shantytown's president quickly intercepted the two snooping gentlemen. He introduced himself as Herr Schmidt, a tailor. Schmidt was clearly used to wealthy strangers showing up in Barackia. The place had become a destination for elites seeking adventure in Berlin's first wave of ghetto tourism. Schmidt invited Ring into his home, where one room served as parlor, bedroom, dining room, and kitchen. Ring was surprised at the comfortable furnishings, the lace curtain in the window, and china teapot on the field stove. Schmidt's wife sat at the table nursing a newborn. She said little, but glowed with the quiet transcendence of a new mother.

Ring marveled at the infant: "The young citizen was enjoying the best of health even though he had been born in a shack." Ring's astonishment was fitting. As Berlin's population spiraled in the last decades of the nineteenth century, babies died at alarming rates, infant mortality peaking in Berlin's working-class neighborhoods at 40 percent. Schmidt and his wife commemorated their liberation from the deadly diseases of Berlin's filthy tenements by christening the boy Freifeld (meaning "open-field") Schmidt.

During momentous events, people often dream up new names to mark what they see as a change in the social order. The hut, the garden, the tightly bound community where residents feared neither thieves nor epidemics, a spot so pleasant it drew curious upper-class tourists, pried open for the young parents a porthole onto a future more promising than life in Berlin's dark tenements, which reeked of piss, rot, and failure. Hope, in other words, comes in many forms. This one took shape lying in his mother's arms: Freifeld Schmidt, born in a shanty in a meadow on the edge of the world's fastest-growing city.

The year before, in 1871, Prussia won the Franco-Prussian War. Otto von Bismarck used the victory to merge principalities into a unified Germany. He financed Berlin's transformation into a world capital with some of the five billion francs in French war reparations. That money flowed into Berlin to refashion the sleepy garrison into a boomtown. Journalists compared Berlin to American frontier cities of big shoulders, steelworks, railroads, and slaughterhouses. In Berlin, chemical engineers came up with novel technologies to process imported food arriving from around the globe. They pureed peas to make plant-based sausages with a long shelf life and transformed palm oil into margarine. The Schering company concocted new drugs, chloroform and cocaine, to overcome pain and exhaustion. You can see the trajectory of these inventions; cheap, nonperishable food and numbing drugs to leaven extended working days in dangerous, monotonous service in factories and armies.

As Berlin became the vortex of German industry, the city's wealthy classes went on a spending spree. They started new businesses, speculated in real estate, and built palatial villas. The kaiser financed a parade route of monuments through the Tiergarten, Berlin's famous counterpart to Central Park. In this extended victory lap, no one thought about the city's poor or lower middle class, although they were a growing contingent. Several hundred thousand migrants arrived in Berlin each year.

As almost everywhere in Europe, changes in land tenure enclosed common land that had formerly supported working people. In Germany, the high price of improved agriculture and drained wetlands east of Berlin, in former Polish territory, drove peasants and craftspeople out of jobs and from ancestral homes. Families piled their furniture, war medals, and gold-trimmed plates with the image of the kaiser onto dog carts and horse-drawn buggies. Migrants went abroad or headed for the new German capital. The newcomers spoke a version of German or a version of Polish, which eventually melded to become a distinct Berlin patois.

Prying working people from their land and means of subsistence loosened the constraints on human reproduction. In Germany's eastern territories, the population shot upward from 1840, after the first decades of enclosures. For a long time, demographers assumed that prosperity and better hygiene led to the period's riotous population growth. But the reverse is just as often the case. As in England, new farming schemes knocked rural people from their homes, lowered the age of reproduction, and the population multiplied beyond all historical precedent. By 1874, nearly a million people lived in Berlin, most of them born outside the city in the rural, eastern territories of Prussia.

There was no room for all these people. In the crush, rents rocketed upward. City officials, Prussian bureaucrats, even the kaiser himself spoke of the "housing crisis," but they all agreed that it was not a problem for the state to fix. Free markets, they said, would resolve the crisis. Developers threw up large tenements that occupied most of a city block. Tenants handed over a third of their income for small, damp living quarters. Trying to make rent, families sublet a room or just a bed to strangers. In the mid-nineteenth century, Berlin transformed in just a few decades from a small city into the most densely populated metropolis in the world.

There were handsome profits to be made. Landlords leased properties for three months at a time so that they could raise rents frequently. As rents catapulted, Berliners also paid higher taxes. Taxes went to the police, who aided landlords in evicting tenants. In 1872, fifteen thousand Berliners, nearly 2 percent of the population, had no place to live.

Herr Schmidt, the shantytown president, took Ring and his friend on a tour of Barackia. He explained that he had rented a large city block from the municipality and divided it, subleasing plots of about

185 square meters to families who built huts and dug subsistence gardens. After 1862 other "colonies" like Barackia sprang up just outside the city's medieval walls. Gardens in these shantytowns were important because food prices were the highest in a century. Over a decade, thousands of people took up residency in the "Barrack Republic" and other garden shantytowns surrounding the medieval walls of Berlin. There, with affordable rents and growing a portion of their food, families could relax a bit. Schmidt's colony had a little shop with a window display of oranges and gingerbread. A café sold beer and meals of pork and sauerkraut. Ring noted a couple of men sitting at café tables deep in conversation, discussing, "with that special German thoroughness," the "social question"—how the wealthy profited from the poor.

Schmidt led his guests past a young potter digging an extra well for the colony and a blacksmith casting zinc columns on a forge in front of his shack. Another journalist observed this do-it-yourself mentality in Barackia: "through their craft and industry, the community provided for its basic necessities."

Ring watched a man plaster his hut, an illegal act. By police order, structures were to be "arbors" with transparent walls. But the police need not have bothered with this code. Poor Berliners in crowded housing spent most of their time outside on the street. Their poverty already made their lives transparent to police and elite critics. For this reason, historians know a lot about working-class men who beat their wives, about alcoholism, prostitution, and other social problems among poor people. The wealthy and professional classes appear, in retrospect, to be more virtuous because they could hide their transgressions behind thick walls.

Schmidt and Ring paid a visit to a rag merchant. A couple and their children sat on sacks and sorted a large mound of cloth. The family made a good living in the rag business, but they could not afford Berlin's rents, nor did they want to live in a dank cellar while sharing an overflowing outhouse with dozens of others. Not when they could

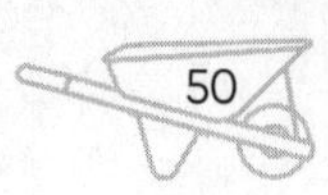

have their own house, garden, and privy. As another resident, a weaver, explained to Ring, "I'm supposed to pay 90 Thalers for a cellar hole, where here my boards [to build a hut] cost 24 Thalers and the land just three. For that, I live like a prince in healthy air and don't have to be beaten or cheated by the landlord."

"But what will you do when winter comes?" Ring asked.

"We will cover the boards with turf and tanbark, nice and warm, and the flies will leave us in peace."

On top of textiles, the weaver knew something about building materials. Peat turf absorbs moisture. Bark holds in heat. Scents from both materials repel insects.

Ring compared Barackia squatters with American settlers colonizing the Great Plains in the same years. He celebrated shantytown residents' rugged individualism and independence in contrast to the Prussian state's overbearing paternalism. German liberals at the time sponsored paupers' gardens, much like English allotments, where under the guidance of an overseer, poor families grew their own welfare. These "Schreber gardens" or Kleingärten (small gardens) had much in common with their English counterparts. Reformers, worried about the moral and physical degeneracy of the urban workers, sponsored allotments to rebuild soul and body through fresh air and fresh vegetables. The Schreber and small gardens came with many rules designed to instill a work ethic, sobriety, and piety. Factory owners sponsored them as a way to skimp on wages by using women and children to grow a large portion of the family's groceries. The unplanned and self-governed Barackia, later called in Berlin "arbor colonies," differed fundamentally from the charity gardens, but were they really similar to homesteading pioneers?

Hmm, maybe not. Homesteaders won deeds to territory that the US Army wrested in genocidal campaigns against Indigenous Americans. Once pioneers proved up, they had all the powers private property granted them to manage their land and exclude others, violently if necessary. Berlin shantytown dwellers rented municipal lots until an investor willing to pay more for the land came along. They had short-term leases and no rights. Their huts and gardens served as placeholders for future development. Although Barackia residents were happy to be liberated from landlords and tenements, they did not have permanence, independence, or the right to exclude others. Barackia residents had already lost their homes and village commons. They understood what homesteaders from Kansas to the Dakotas would come to know within a few decades: The security of private property is nothing to bank on, so easy is it for working people to be deprived of their land, homes, and means of production.

Instead of property-backed, rugged individualism, Barackia residents relied on each other. Households lived together in small colonies of 150 to 300 people. They managed by consent the governance of an informal community that handled small industry, enterprise, and the construction of infrastructure. Colonists built their streets, fountains, and public spaces. They solved the problems of shelter, waste, hygiene, and food security with no public funds, no planning or bureaucracy, no elite leadership or intervention. Tourists and journalists were drawn to Barackia because it offered an alternative to the wretched slums of the emerging megalopolis. Ring celebrated the workingman's paradise he had discovered in the "Free State of the Barackia," "a state in the open air, on an open field, with free ideas and institutions, spared meddling police and tyrannical landlords, liberated from rents, taxes, polluted gutters and unsavory cesspools, free from all the burdens and torments of the cosmopolitan city."

Not everyone was quite as gushing as Ring about the unregulated shantytowns. A journalist for the *New Social Democrat* described thrown-together shacks with smoke pouring from cracks in the walls. He noted the humiliating lack of privacy, but he also allowed that Barackia had a certain *Gemütlichkeit*, or coziness, with children playing in the sand, elderly folks weeding the gardens, and in the evenings after the men returned from work, chatter, laughter, and "everywhere the sound of accordions." Like Ring, this journalist called Barackia an "idyll."

Like a jar of pickles, daily life ferments into culture. Culture on the most basic level brings together disparate ideas, practices, and living beings to brew into something novel, satisfying, and sometimes miraculous. Barackia's musicians, storytellers, and informal news gatherers nurtured a working people's civilization, which took shape as they sat before their huts in the evening, the murmur of voices and accordion music tracing the outlines of the community.

When Ring visited in early summer 1872, the colony had been in place for ten years. The people of Barackia appear to have believed they had solved their housing problems for the near future. Ring did not report that anyone mentioned that their days in Barackia might be numbered.

Pablo and Maria started gardening in May just as Washington, DC, was heating up. They spaded up the turf alongside a branch road in Rock Creek Park, a tree-filled, green thread that winds through the District of Columbia to the Potomac River. I wondered how long it would take the park police to notice the couple's illegal garden on national parkland. The day I stopped to talk to them, Maria, in her late forties, wore a pink, satin tracksuit; Pablo had on cutoffs, a tank top, and good boots. Both had the weathered look of people who spend most of their time outdoors. They spoke only a little English. I only a little Spanish. Our exchange quickly trailed off. They were carrying buckets of water from the stream to their little plants that were struggling in heavy, red clay. I picked up a bucket and started hauling too. We smiled at each other, heaved, and poured.

After a while, my son, Sasha, who knows Spanish, joined us. He interpreted. Maria explained that they were digging in the park because they loved plants. They had grown up on farms. She was from Mexico. Pablo from Honduras. Maria showed me her zinnias, just blooming. The couple slept on the front porch of an abandoned house and spent their days under a shade tree in their garden. They had no jobs. Neither of them had been to school, Pablo said, "Not one day." They had lost track of their families back home. After two decades in the United States as undocumented migrants, they had a couple of bags of clothes and this little borrowed plot to call their own. That's all.

Maria said that if they had a hose, they could use gravity to draw water from the stream and save a lot of hauling. I kept an eye out and retrieved a discarded hose in an alley. When a few weeks later I headed to their plot with the garden hose, I found a not-so-surprising scene. Two park police officers were there to evict Maria and Pablo, who were stunned and upset. Who else was using that piece of land? they asked. They looked in despair at the plants they had cultivated. The corn was just under the knee. Flowers poked up here and there, looking lonely among a few sprouting beans. It was by no means a lush garden, but gardens are like babies. To the parents they are beautiful. We stood about mournfully, not knowing what to do.

Then suddenly the situation changed. A man showed up driving a pickup—a white guy who spoke Spanish. He explained that he had found Pablo and Maria a plot in a community garden. They just had to move their seedlings to the new place. We got to work, digging up and placing the plants in plastic pots and bags and stashing them in the bed of the truck. Pablo and Maria climbed in, too, with their little green charges, waving to me happily as they drove off.

I didn't see the pair again until a year later when I discovered them in a cloverleaf patch of land in the middle of a major artery east of my neighborhood. Amid the swishing cars was a green space that had been fashioned into a community garden. They were sitting in front of a tent. We had a short reunion. Using words and gestures, they told me what had happened since I last saw them. The man with the pickup truck had taken them to a community garden, where they transplanted their seedlings. Then he drove Pablo to his house and in the ensuing days put him to work, drywalling his basement and attic, tiling, and framing. "I worked and worked," Pablo said, "morning to night."

"Did he pay you?"

"No, he gave me food and a bed. He said he was going to find me work for pay. He told Maria to go." [finger pointing] 'Get out of here!' "

Pablo's eyes widened. "He said that to Maria! Maria and I are together."

After two weeks, Pablo left. The couple moved into the community garden, where they slept near their plants for a short time, until the community gardeners told them to leave. Maria turned to me and gravely enunciated something she wanted me to know. When evicting them, the community gardeners had chopped up their plants. Her hand moved up and down as if wielding an axe.

Maria and Pablo left dejected and defeated. But there is always another garden. They moved to the Wangari Gardens, where we sat, a community project dedicated to the Kenyan environmentalist Wangari Maathai. The garden had a grove of fruit trees, offering shade and privacy. It had rain barrels, a toilet, and a welcoming community that gave them a tent and invited them to stay in the garden as caretakers. It was a good decision for the garden, which looked well-watered and weeded.

For Pablo and Maria, they were reunited and had a place to stay, but they lived alone in the middle of a traffic artery in a tent with no heat or electricity. They were often too cold or too hot. One day, someone broke in and stole Maria's clothing. "Only mine," she said in anger.

The couple lasted less than a year in the Wangari Gardens. After that, I lost track of them. I wondered how their fate would have differed had they landed in a community of unhoused gardeners as did the parents of Freifeld Schmidt, where along with their plants, Barackia residents cultivated mutual aid. If, instead of living as isolated homesteaders, they had squatted in a community like Barackia, would they have suffered less from crime, exploitation, and loneliness?

Maybe. If history is our guide, let's follow it.

Rather than liken Barackia to mythical American frontiersmen, Ring could have compared it to the Paris Commune, an experiment in self-government that occurred a year before Ring toured the shantytown.

The Paris Commune came together at the end of the Franco-Prussian War, amid a dire food shortage. As the French army lost control of Paris, in March 1871, rebelling workers and disaffected soldiers took over the city and quickly established a new order. The revolutionaries set up a commune and passed legislation aiming for a more just society. They called for rent remission, abolition of child labor, debt reform, a ten-hour workday, self-policing, separation of church and state, and employee-owned businesses. The Paris Commune took over the city's tax revenue, handed out food and clothing, and set up free schools and hospitals. A women's union called for equal wages and rights. In long, contentious deliberations, Parisians for the first time made decisions that shaped their immediate world.

The Paris Commune lasted only seventy-two days. Parisian communards commandeered cannons and built barricades, but they were no match for the battle-hardened French army, which did not hold back in crushing the social experiment. In the aftermath, thirteen thousand captives were sentenced to execution or imprisonment and ten to fifteen thousand communards were killed in the fighting. These were exceptional mortality rates for the time, bloodier than the Battle of Gettysburg.

A year later, Berlin's working poor also sought relief from low wages, speculative markets, and unrestrained exploitation. Thousands attended meetings in the summer of 1872 to draft proposals for a ten-hour working day, living wages, a mass tenants' union, and a return of common land to the people, to be administered by the people. Berlin's spontaneous, unplanned shantytowns offered a vision of the kind of communal, democratic governance and mixed economies that working people built when left to themselves. They offered a prototype for a kind of proletarian urbanism that had been crushed a year before in Paris.

With such anarchic movements swirling about, German authorities worried a great deal about social order. In 1872, the newly established "building police" sought to dissolve Barackia and other informal settlements. But there was nothing illegal about the garden shantytowns. Residents had leases and a right to occupy the land. Their primitive dwellings and gardens violated no regulation because Berlin had no building code at all until January of 1872, when city officials ruled that builders must have construction permits. No one in Barackia had permits, but most had built their huts before the code.

Even so, the building police posted a notice on July 26 outlawing unpermitted dwellings in the settlements surrounding the city. With this news, agitation spread. Where would people go? At this time, real estate speculators were buying and selling the same properties up to twelve times a day, each sale earning the seller a tidy profit, while rents levitated far beyond wages. The leftist newspaper the *New Social Democrat* warned that the housing crisis was so acute and "landlords so arrogant" that any spark could send the city into conflagration.

And on July 27, a landlord kindled a flame. The police were going about a routine duty, evicting an elderly street peddler, Ferdinand Harstark, from a tenement apartment on Blumenstrasse in Berlin's center. Harstark, "an invalid," resisted ejection from his home where, he explained, he had lived all his life. He showed the police his lease and receipts, proving he had paid rent on time. But the landlord had found someone willing to pay more, and he wanted Harstark out.

Seeing Harstark's furniture and bedding disgorged on the street,

Berliners moving house. Hirtenstrasse 9, 1901.

neighbors gathered to defend him. Soon a crowd of grim-faced people formed. When the shift ended at a nearby factory, the workers joined the throng. Young men broke the windows of the landlord's building. The disturbances drew more people. A small police force on horseback trotted in, but by evening four thousand enraged people stood on the street, and the horse guard was no match.

The commander called for reinforcements and announced that law enforcement would clear the street at nine that evening. He cited a much-hated police ordinance that criminalized standing on public sidewalks. But that is what Berliners did when they had free time. Because their homes were dreary, they tarried outside to visit, gossip, and gather news while taking in light and air not supplied in airless apartments. The crowd reacted violently to the order to disperse. A cascade of stones poured down. Police pushed the crowd, and the crowd shoved back. A melee ensued. In the brawl, thirty police officers were injured. No count was made of wounded civilians, but police sources in the Berlin city archive noted that women and children were sent to the hospital. It took authorities most of the night to remove the protestors and finally evict Harstark.

The next day, unrest spread further. Demonstrators climbed streetlamps and turned the gas lanterns into flaming missiles. They tore up cobblestones from the streets to make projectiles. They lifted feces-caked planks covering gutters and used them to build barricades.

Kaiser Wilhelm I wrote a letter to the minister of the interior demanding that the "excesses" be met with all possible force so there would not be another revolution the likes of which had quaked the Prussian Empire in 1848, nearly toppling the monarchy.

For three days, police and two battalions of the Emperor's Grenadier Regiment fought the working people of Berlin. Hundreds of officers armed with sabers and pistols went into combat against thousands of citizens wielding rods, stones, knives, and flowerpots. A police reporter called the street fighting "a formal guerrilla war." During three hot summer nights, wherever police turned, angry crowds attacked.

On the fourth day, storm clouds crowded the horizon and billowed upward. Great gusts brought a drenching rain. Because of the weather, the police chief held his forces from the streets. With no officers to battle, the flames of disturbances died out. Later in the week, courts sentenced eighty-five people to long prison terms. Only eight of the convicted, a liberal newspaper reported, were Berliners. The rest were "outsiders" who, they claimed, had no right to residence in Berlin. "Outsiders" were recent migrants to the city. Just by dropping a few words into a sentence, elite Berliners reclassified the city's public spaces into exclusive reserves for native-born "Berliners." This is how territory gets defined.

During the unrest in the city center, peace reigned in Barackia. The shantytown had no streets, no rent hikes, and so no police and evictions. But a few weeks later the police chief announced a plan to raze the city's informal settlements by mid-September. Those who had not found other housing by that date would be taken to the municipal workhouse. No one wanted to go to those asylums for the poor, "houses of horror."

Panicking, a delegation of Barackia residents went to see the police chief, Guido von Madai, who dismissed them. A group of women made a visit to Mayor Arthur Hobrecht, older brother of the master architect of Berlin's city plan, James Hobrecht. Depending on who reported the news, Mayer Hobrecht either promised they would have a stay of eviction until more housing could be built, or he shrugged them off, busy public official that he was. A shoemaker, Albert Haack, wrote a desperate letter to the kaiser, begging for a postponement of eviction until October, when their leases legally expired. He signed his letter with the only address he could conjure: "shantytown in front of Landsberger Gate, second row, first booth." The kaiser did not reply.

Instead, an order went out to pull down Barackia. It appears the timing of the demolitions had a lot to do with a September meeting of three monarchs that the kaiser was hosting in the new German capital. It would be unseemly if the Russian tsar Alexander II's first glimpse of Berlin was shantytowns. And there was money to be made from housing shortages. Half of Berlin's city councilors were landlords. For them, overcrowding and high rents padded their fortunes.

Police officers had learned a lesson from the July uprising a few weeks before. This time they would not make theater in broad daylight of the savage act of dispossession. So, after midnight on August 29, a phalanx of police and soldiers on horseback and foot marched into Barackia. Policemen in blue wool circled the huts. A commander ordered sleeping residents to show their faces. People stumbled outdoors. Children rubbed their eyes and shivered. The families, in nightclothes or underwear, watched as firemen dragged out their still-warm bedding, dishes, furniture, clothing, sacks of flour and potatoes. They dropped them in the dirt. With axes and sledgehammers, the fire-

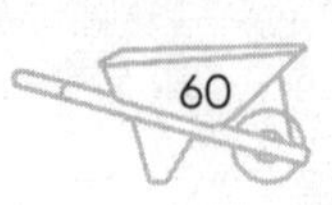

fighters turned to the flimsy dwellings the Barackia families had built. Hammers snapped the ribs of plaster-lathe walls.

I imagine the mounting rage a fireman felt as he perceived in the gloom the eyes of residents watching him. He, too, had trouble paying rent. A trip to the grocery store often erupted into an argument. His wife said the price of everything was rising. Maybe that was true, but his family made do, while these people here—who were they, so helpless and pitiful? Why did they just stand there? The fireman picked up his axe and directed his fury at the shantytown residents, their vulnerability and implacable fatalism.

He smashed away at the domestic comforts of Barackia as if hacking away a stand of weeds. The authorities destroyed garden vegetables valued at a year's wages. They razed workshops and so demolished the future earnings of Barackia artisans. As the police and horses stomped on tomatoes nearly ready to harvest, the scent of the crushed plants wafted along in the darkness, the red-fruited nightshade registering the eviction with an aerosol alarm.

Those awake on Kottbusser Damm Street late into the night of August 29, 1872, witnessed a strange spectacle. Under the gas lamps, a ragged parade passed into and out of each pale, blue spotlight. A grim tableau. Half-dressed men holding rabbits and goats shuffled alongside women in torn petticoats leading toddlers by the hand. Public officials in peaked helmets followed. Drivers poked at teams pulling police carts mounded with pine tables, oak bureaus, wall clocks, and crockery. The clack of horseshoes on stone echoed against the row of townhouses. This tangled mass of humans, animals, and materials proceeded along the grand boulevard toward the workhouses on Alexanderplatz.

As they went, I imagine the wail of little Freifeld Schmidt trailed after them.

The Working-Class Origins of the Garden City

ON A WARM JUNE DAY IN 2023, Rosa Oefinger and I went looking for the place where Barackia once stood. Rosa worked with me helping to find archival documents in Berlin to piece together the history of working-class gardeners. Rosa had the idea of retracing the journalist Max Ring's steps along the canal to the former "Butcher's meadow" outside of the Kottbusser Gate, in what is now the Kreuzberg neighborhood.

Berlin on a Saturday afternoon in June was all bustle and noise. Advertisements flashed by on elevated trains. Bikes, cars, buses, and joggers hustled past us, spinning a kaleidoscope of color. Berlin is not a cozy city—by design. In the 1860s, Berlin's first city planner, James Hobrecht, plotted out wide, straight streets, prioritizing traffic circulation for commerce and industry.

Following Ring, Rosa and I sought quiet in the dappled light of the trees hovering over the Landwehr Canal. Rosa had brought copies of maps of Berlin from the 1860s to the 1890s. We walked and studied the maps, seeing how quickly, from decade to decade, Berlin outgrew the confines of its medieval walls when developers punched holes into the boggy heaths, sandy wastes, and meadows surrounding the city to make room for buildings. In 1870, the edge where we walked served as a staging ground for grander parts of the city. We passed the relics of a gasworks, a pauper's graveyard, a home for juvenile delinquents, a large garrison for soldiers, and a few breweries. Between the streets and boulevards, in Ring's day, were vacant lots waiting for development.

We strolled into the heart of Kreuzberg, where Berliners sat sunning in outdoor cafés before sweating, orange glasses of Aperol spritz. Young people reclined on grass along the canal drinking beer, strumming guitars, or playing keystrokes on their phones. Kreuzberg was gentrified after the fall of the Berlin Wall from a hard-to-reach immigrant neighborhood into hipster-central. Here Rosa and I studied the maps more intently. Nineteenth-century mapmakers did not include precarious housing in their drawings. Barackia, we realized, had not been right on the canal, a busy traffic intersection at the time, but set back a hundred meters. We saw that a hospital was built on the site, construction starting a few years after Barackia's demolition. We could see a towering new hospital built in the 1960s, but where was the old one from the 1880s?

Our maps pointed us to a narrow corridor between two brick buildings. We stepped into the passage, walked through it. Tumbling out the other side, I felt like Alice falling down the rabbit hole into Wonderland. We were on the grounds of the former hospital. And such grounds. The old hospital pavilions were now apartments opening onto a courtyard. There, the noise and crowds evaporated, replaced by birdsong and the hum of voices in conversation. Around us botanical life luxuriated. Vines climbed walls. Roses, verbena, lilac, mullein, serviceberry, and pear trees extended their limbs to channel the sun. Two women were setting out a table and an umbrella. They had a teapot and plates of *Apfelkuchen*. A young man read a novel on the grass. Several children jumped about in a game to which only they knew the rules.

The 1880s hospital was comprised of a series of pavilion-style, two-story wards built out of red and yellow brick, linked by courtyards. The hospital wards and gardens worked together. The idea was that patients needed a daily dose of sun and fresh air in a garden setting to heal. At the time, one of the most feared diseases was tuberculosis. The only cure was for patients to rest and take in fresh air.

Rosa and I, worn out from our walk and the rumble of the city, looked at each other in the cool shadows of the plant-filled courtyard. We smiled, reading each other's thoughts. We had found Barackia.

Over the decade that Barackia residents had gardened at the Kottbusser Gate, they had regenerated desolate territory churned up by drainage projects, farming, and construction. Architects looking for a place to locate their hospital and gardens would have come across the Barackia demolition site. I imagine their surprise, kicking aside discarded rags and rotting boards, at finding rich, dark hummus, and a full complement of plants: perennial sage and thyme, strawberries, raspberry canes, and fruit trees. Barackia's rare verdancy would have been a selling point for a garden hospital. Capitalizing on the gardeners' labor would be a cost savings in landscaping the hospital grounds. It would not be the last time that gardeners' floral effluence inspired real estate development.

In a ghostly fashion, Barackia, demolished 150 years ago, lived on as a tenacious green shadow, and, more generally, as a set of practices and a form of organization that persisted after the shantytown fell.

Elites in the last decades of the nineteenth century feared urban working people as their numbers grew in cities. Wealthy urbanites were outnumbered by working people ten to one, yet they depended on masons and carpenters to build their homes, on factory workers to manufacture goods to furnish them, and on servants to polish and plate the luxuries. Dependency bred contempt and nervous glances over the shoulder. Kaiser Wilhelm II called urban workers a "gang of traitors, a breed of men who do not deserve the name of Germans." Journalist Victor Tissot went on his own ghetto tour of Berlin a few years after Max Ring. He came away appalled at what he classified as an "invasion of déclassé, adventurers, beggars, and vagabonds" camping out in hordes who preyed on Berliners, kidnapping young girls, robbing and murdering with no remorse.

Western culture is focused on texts—laws, constitutions, manifestos. People often missed ideologies voiced in actions by hands working out ideas with and through materials around them. For example, real estate developers and the Berlin construction police wrote manifestos onto the landscape. They replaced soil and dirt with stone and asphalt. They swapped out trees for streetlamps. In the years after the discovery of germs as the source of disease, authorities sought to eradicate earth, animals, and bacteria from city spaces. These manifestos drew a line between city and wilderness, between clean, orderly urbanites and the dirty rabble.

Yet, working people possessed their own texts. Shantytown dwellers rejected the warehousing of humans in dark tenements. They turned their backs on the hierarchical residential structures where the poor lived in moldering basements while the wealthy lived above, commanding the light and air of the street. Despite Barackia's dissolution, more and more people in the following decade built garden shantytowns along the swelling edges of Berlin. The refusal to live in squalor was another manifesto, powerful and persistent, written onto Berlin's landscape.

Historian Hartwig Stein argues that spontaneous garden shantytowns or arbor colonies, as they later became known, sprang to life anew in the years after Barackia's destruction. Elites set up Red Cross, charity, and pauper gardens focused on improving working-class health and morality. Workers often rejected them.

Arbor colonies were different. Gardeners organized in associations, signed leases, and elected five-member committees for leadership. Members would clear a vacant lot of junk and rubble and name their settlements descriptively—Stony Ground (Steinreich) or Sahara—after the rocky, sandy terrains they took over. With the introduction of the eight-hour working day in 1895, people had more free time. They scavenged the city for organic matter and perennial plants and foraged in dumps and building sites for scraps to build cottages. Everything had a second life. Stones dug out of gardens became foundations. Branches and reeds made fencing and trellises. One family repurposed a train car, attached a tiny tower to it, and called it a "castle." They dug wells and shared seeds and grafts with neighbors. At the turn of the century, as they worked, the number of orchards and small livestock in industrializing Berlin rose.

Working-class gardeners had little money for artificial fertilizers. Instead, they constructed compost piles and tossed in what others threw away. "Everything belongs in the compost heap"—brewer's mash, slaughterhouse bones, tanners' barks, kitchen scraps, ashes, dishwater, animal manure, and the rich human waste from under the privy. They piled up this rummaged bounty, got out their spades, and turned the compost heap, which would heat up and ferment as ravenous organisms took the piles apart. When gardeners closed their beds for the season, they blanketed them with this compost, which would decompose into a fine, dark tilth by spring. While wealthy farmers in the countryside were using resources shipped from abroad to fertilize their exhausted fields, allotment gardeners created soils from urban waste. They took garbage and turned it into a cornucopia. Drawing on the discards of a large city, the arbor colonies spread across Berlin.

Working people rented on the worst possible terms. Short-term leases stipulated that the middleman retained all improvements from clearing, building, well-digging, road-building, and planting. They also kept the exclusive right to sell building materials and run beer gardens and canteens in arbor colonies. With no competitors, the lessor charged high prices and raised the rents after gardeners improved the property. Landowners evicted tenants who did not spend enough money at the pub. From these depredations, a great antagonism festered between workers and property owners. A police report from 1899 noted the sharp knife of animosity dividing Berliners by class, a "gulf that separates the workers from the rest of society."

When the time was right to build, the landlord or property manager would terminate the lease, and the police would aid in evicting gardeners. Displaced people would pick up and move to new ground, sometimes taking their huts with them. They would rent new lots and divide them into garden realms, and name them serially—Gemütlichkeit (Coziness) I, II, and III. Flexible and persistent as the climbing collaboration of bean on corn, arbor colonies took shape over and over, forming, growing, inventing, and then dissolving in evictions, only to move on.

"We eat salad, yes, we eat salad and eat vegetables early and late," a working-class song went. The urban gardeners were not, like their fashionable social betters, self-proclaimed vegetarians and nudists. Rejecting industrialization, the nineteenth-century German elite, calling themselves "Life Reformers," flocked to parks where they shed their clothes and ate vegetables to restore physical and mental balance. But working-class gardeners ate vegetables because that is what they had. Nor did they invest much in the notion that gardening taught discipline and a work ethic. As one gardener put it: "We don't want to toil too much. We make it easy, helping each other out. We must not be slaves to the earth, but freelancers with time for leisure."

By 1900, fifty thousand families had garden allotments either in Berlin's self-governed, arbor colonies or in the charitable Schreber or Red Cross Kleingärtens. Arbor colonists created little worlds within the big city. As needs came up, they organized amenities to service them. They built playgrounds, libraries, clubhouses, summer camps for kids, and dance floors. In June, gardeners held a summer celebration. In fall, a harvest festival would start midday with a parade, a brass band, young girls in white dresses holding wreaths, and children running after an uncle tossing candy. Someone grilled bratwurst and boiled potatoes, the hot dishes piled on tables with pickled garlic and mushrooms. The gardeners invited elderly neighbors to tea and cake. Visitors went home with fruits and vegetables, surplus from the gardens, and were encouraged to make compote and pickles before the produce went bad. At dusk, the band started again and there was another parade. After the children were marched off to bed, the adults tarried, dressed in their best, dancing. The heavy stomp of the tuba beat a waltz while the dancers spun. The ring of the trumpets levitated over the rows of huts, the bright, metallic notes gathering up this bit of the huge city, making it familiar, home.

Summer and fall, life in the arbor colonies was best. But winters in the small, primitive dwellings were a trial. Berlin sits low on the landscape. A granite-gray fog creeps in. Continental winds peel across the plains, bringing bone-rattling chills. In cottages with thin walls, the draft blowing through, children suffered frostbite and burns from stumbling into iron stoves. Arbor colonists made improvements at great risk. If they paid for a foundation or electrification, they lost their investment when the property manager did not renew the lease. Colonists found solutions to the housing crisis by building their own communities, but they did not have the power to assure any kind of permanence. Faced with the exclusions of property and the riotous gyrations of markets, more and more people began to question why housing and food sovereignty were not fundamental rights.

In February 1901, thousands of Berliners got together in a mass meeting and voted to create a union of planters' associations, which published a news bulletin, the *Ackerbürger* (the field citizen), with the aim of bringing "light, air, free association, warmth and friendship to arbor colonies." Initially, only a few colonists joined the union because property managers threatened to evict gardeners who signed up. Fighting the hold of landowners, the planters' union turned to politics. They demanded permanence in the form of thirty-year leases and sought to situate garden communities on public land to escape the grasp of eviction-happy landowners. They formed cooperatives for wholesale purchases of coal, potatoes, plants, and seeds. They bypassed landowners by pooling savings to buy land. Gardeners petitioned to have postal addresses for mail and for the right to send their children to school. In 1908, socialists in the organization voted to impose a general ban on beer gardens. That freed up arbor colonies to operate canteens, which circulated profits back to the community. They took up collections for workers' compensation, life, theft, and fire insurance so that when a member's goat was stolen, the association paid to replace it. They started courses for gardening and animal husbandry, set up arbitration courts for disputes and sanatoria for tuberculosis sufferers. Through the planters' union, gardeners shored up their precarious existence.

On the defensive, landowners and leaseholders established a rival union of their own. Developers, architects, and city planners called the arbor colonies "grotesque" and unclean, hotbeds of moral depravity and political radicalization. Yet, even so, something was appealing about them, for all their primitive qualities. Reformer Max Christian put it this way in 1914: "According to unanimous reports, there is great solidarity in the garden colonies, a solidarity which moves in fixed forms without a written code of honor. Unfair elements are soon recognized and branded. Almost everywhere there is a competition as to the beauty of the gardens, the originality of the arbor houses, or the richness of the harvest. In joyful celebrations, which are common almost everywhere, the spirit of community blossoms."

These were the years when workers were going to meetings, debating, lecturing, learning, becoming mobilized in clubs and organiza-

Before and after: An unimproved 1924 arbor colony in the Rehberge region of Berlin.

A lush arbor colony in Neukölln where gardeners have regenerated rocky, sandy ground. The gasworks are in the background.

tions, when the Social Democratic Party won the majority in Berlin's city council and a third of seats in parliament (which had no real power). These were the years when the cost of basic food staples rose as the kaiser built up the German navy, igniting inflation, the same years when Berlin remained a battleground between police and workers.

But the arbor colonies were different. The narrow earthen paths belonged to the people who created them. Police had trouble orienting themselves in these informal zones and tended to stay away. Housing authorities tried to take an inventory to register who lived where to stop the sprouting of illegal dwellings and prevent "the gathering of criminal riffraff," but they rarely succeeded. In subsequent decades, the Weimar and Nazi governments tried and failed to conduct a census in the allotment colonies. Bringing a statelike order to the colonies was like trying to hold onto a handful of sand.

As the colonies hummed away, plants spread over neglected Berlin wastes, often former fertile wetlands that had been drained into barren sand dunes. In photos of arbor colonies, two things stand out. Some new or temporary gardens are hard to look at, with broken-down sheds, bald parcels, weeds, and plants struggling in Berlin's sand. But colonies that persisted for more than five years flourished. In photos, shrubs and vines crawl up and over modest dwellings, nearly obliterating the buildings from the camera's eye. Plant life, more than human-built structures, formed the infrastructure of these urban neighborhoods. The plants sent roots, sensitive and searching, under the terrain of daily life, feeding, shaping, and enhancing it.

In 1898, the English socialist Ebenezer Howard dreamed up the concept of a garden city. He imagined neighborhoods where industry, agriculture, small dwellings, and gardens would come together on communal land to address the problem of the crowded metropolis. Howard's book caused a sensation in Germany and around the world. Yet shantytown gardeners were way ahead of him.

Leaf Town

LOOKING BACK YEARS LATER, people who lived in the green shantytowns remembered not drafty walls, leaking roofs, or outdoor toilets, but (here is that phrase again) "an idyllic landscape," children loping about freely at play, women catching up over the hedge, adults at work in gardens and shops, and fresh produce, grown and traded, weaving a web of social relationships. Surely it wasn't all a rose garden? Memories work as filters, straining out painful, banal, and unpleasant moments from the past. What were the bad parts that people forgot?

There were plenty of conflicts. Gardeners complained of others who did not care for their plots, who let weeds run and windows remain broken. Newspaper reports record the murder of a landowner in an arbor colony, and of a man bitten to death during a drunken melee at a harvest festival. During hungry times, people had to bring their chickens and pigs indoors at night, lest they be stolen. Some

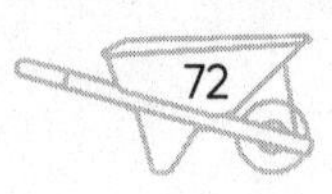

people got drunk in the beer garden and kept other people up with their songs and quarrels. People fought over politics. I assume lots of people got sick of their small quarters and their neighbors and moved somewhere else. But the garden settlements persisted. For a century from 1860 to 1960, Berlin could not do without them.

Rosa and I took a walk in an allotment on the outskirts of Berlin on a sparkling Sunday morning. A couple was having breakfast on their terrace in front of a fine bed of greens and tomatoes. We fell to talking. The woman told us they had lived in an apartment in Kreuzberg for forty years. For decades they paid no more than 400 euros a month for rent, but then the wall came down. Kreuzberg gentrified. Rent climbed to 4,000 euros a month.

"We can't pay that."

"Then, do you live here?"

"No, it is prohibited to live in your garden cottage full time."

I noted the mailbox, the little driveway with a car in it, the spacious little house with propane gas tanks. It was clear they were living there. Looking around, we saw they were not the only year-round residents in the colony.

I understand that the story of people making ends meet by relying on tiny gardens and mutual aid doesn't sound very believable. Critics will respond that people don't just form themselves into self-governing communities, create a kind of commons, reshape their environments, and dig themselves out of desperate straits simply through self-organization and collaboration. Human nature, we are told, doesn't work that way. Society needs a higher authority—a state, governor, king, or dictator to direct unruly people or they will descend into a *Lord of the Flies* free-for-all.

But what if the standard portrayal of "human nature" is wrong—as in, from the very beginning? And animal nature, too, for that matter, from which so many analogies to humans have been made? Before anyone mentions "utopia" and dismisses the history of gardeners creating bucolic worlds in an industrializing city, I will digress a bit into deep history. People often look to the past to figure out how to live. They eat paleo diets and run long distances in imitation of hunter-gatherer ancestors. What does deep history have to say about garden shantytowns?

The story usually told is this. Hunter-gatherers lived in a dark, periodically hungry and violent world where they struggled with both the elements and wild beasts. Once they discovered agriculture, they mercifully settled down and ceased their nomadic (and apparently pointless) wandering. This first "agricultural revolution" allowed for surpluses to feed people who no longer had to spend all their time searching for food. Agriculture enabled people to gather in the first cities. With food and tools stockpiled, some people had time for creative endeavors, but they also had to worry about protecting their stores from marauding barbarians. Kings emerged who took responsibility for the community's security. They built fortifications and armies. They waged wars and took captives who became slaves. Kings and their growing retinue of deputies devised writing and bureaucracy to command others and secure more surpluses and armaments for war. Early agricultural civilizations created lasting institutions like governments, hierarchies, armies, slavery, and taxes. Lower classes and slaves did most of the heavy lifting, while some men and women developed skills to specialize in masonry, metalwork, textiles, and the like. Thus, civilization was born, and states emerged that produced astounding feats of architectural, artistic, and technological ingenuity. Civilization allowed human genius to shine.

There was just one rub. The new states were coercive and shot through with inequality and exceptional violence. Archeologists locate ancient monarchies by finding mass graves that point to mass murder, a reliable sign for archeologists of "state formation." Who would willingly sign up for a society that engaged in routine slavery and ritual murder? Political scientist James Scott argues that people had to be coerced, sometimes sealed in walled cities, to keep them from running away from civilization. Compared to the easy life of hunting and gathering, farmwork was repetitive, dull, physically taxing, and hazardous. Hunter-gatherers were taller, fitter, and healthier than their farming neighbors. They had more free time and autonomy. If our ancestors had a choice, they chose foraging.

But what if this story is not the whole story? Anthropologists David Graeber and David Wengrow argue there was not one trajectory from either cruel barbarians or noble savages to civilization powered by sedentary agriculture. Rather, all over the world over many millennia, the picture was mixed. Human evolution took no one path of development. Some societies were brutal and hierarchical. Others showed no signs of violence. Graeber and Wengrow find a wide range of early and late Neolithic societies that opted not to accumulate wealth, submit to kings, or conquer and enslave others. Still, political theorists argue that, once people started to farm, some form of bureaucracy, private property, and coercion had to emerge for the more complex organization and security needs that agriculture demands. After all, a cultivar is a vulnerable plant that cannot propagate without human help, and so the domestication of crop plants implies some form of domination and a conception of property.

The first human settlements tended to be river-delta societies engaged in what looks to be farming-lite. Flooding rivers brought nutrients to gardens, the seasonal deluge doing the drudge work of fertilizing and weeding. Humans were left simply to poke a stick in the ground, drop in seeds, return to hoe, and wait while the plants grew. This form of agriculture did not lead to private property, because one year's fertile field could be next year's flooded plain. So people held land in common, flexibly shifting garden sites and redistributing plots and pastures as needed. English, Dutch, and German peasants had this kind of system in the marvelously productive fens and swamps of southern England, Holland, and eastern Prussia before they were drained to make agricultural land.

Some of the first cities or "mega-sites" with tens of thousands of residents, such as Nebelivka and Taljanky in today's Ukraine or Çatalhöyuk in Turkey, had no monumental architecture or elite graves—no acropolis, palaces, or great temples. Humans surely had the usual disagreements, but there are no indications of mass violence or war, nor signs of classes of elite and poor. It does not appear that hinterland farmers dragged crops to these cities to feed urban populations. Rather people built comfortable houses surrounding some sort of community center and engaged in small-scale gardening on the edges of their settlements, where they kept small livestock, cultivated orchards, and gathered a wide spectrum of wild foods. Despite a high population density, they don't appear to have been hungry or stressed. They governed themselves, making decisions in public forums. People of Taljanky left behind statues of full-bodied women, which earlier archeologists took to be fertility goddesses but now are interpreted to be likenesses of community leaders. Archeologists speculate that female elders led their communities in improving soil, selecting wild plants for cultivation, and inventing ways to use clay and plant fibers to devise tools and clothing for more comfortable living. These Neolithic urbanites inhabited their cities without draining local resources, which means they did not need to take over other people's territories. In Nebelivka, archeologists postulate that gardeners accelerated the formation of the black earth of the Ukrainian breadbasket. Ukraine's famously rich soils, some of the most fertile in the world, were not a gift of nature but a creation of human ancestors.

So, human history offers no singular course of evolutionary development. It is richer than that, more dynamic, varied, and full of possibilities. Innovations left off in one era were picked up in another. Something similar occurred in the spreading landscape of garden communities in Berlin in a period of fierce industrialization, urbanization, and misery. In the very cosmopolitan bellies of market capitalism, some people inhabited an alternative world resembling a lesser-known trajectory of human history six thousand years before.

Each arbor colony numbered a few hundred people, small enough so that people could know everyone else in their community. The gardeners built no temples, palaces, or prisons, but they did construct cafés, libraries, preschools, dance pavilions, and playgrounds. They governed themselves collaboratively and shared access to land in almost exactly equal proportions. Garden associations managed to mobilize a great deal of labor when needed, without the creation of a class of workers and a class of overseers. Working people in turn-of-the-century Berlin rarely had regular employment. They were on the job, then laid off, then on the job again. Garden produce, chickens, rabbits, and pigs helped them get by during downturns. The gardens, the neighbors around them, the insurance funds they took up not only saved working people from calamities, but they pointed to a more equitable, just, and humane society. Gradually, a few people began to take note of this quiet social revolution on the margins of Europe's fastest-growing cities.

In 1899, Peter Kropotkin published *Fields, Factories and Workshops*. Kropotkin had long been concerned about the ecological transformations occurring around him. Why, he asked, were there deadly famines among wheat growers in Russia and rice farmers in India? While in Paris, Kropotkin observed, five thousand farmers working in tiny urban courtyards grew enough fruits and vegetables to feed two million Parisians with a surplus left over to send to London. How could that be? Why were farmers in the countryside starving, while Parisians could grow an excess of food on small lots in a crowded city?

Kropotkin learned that Parisian farmers were like flies, attracted to Paris's mountains of horse dung. In the early twentieth century, thousands of horses trod up and down Parisian streets. Stable boys swept up the waste, and Parisian farmers bought it for pennies on the pound. The farmers buried fresh manure two feet thick underground, layered topsoil over it, nudged in seeds, and placed a glass window over the mound. Microbes in the manure got to work. As they digested the dung, the soil heated up. The seeds, warmed from below by microbes and above by the sun, germinated in the miniature self-heating greenhouses that the farmers called "hotbeds." In the dung-warmed beds, they planted spring crops in winter and summer crops in spring.

With two tons of horse manure (which would make a mound the height and width of a one-car garage), a Parisian farmer could manipulate the calendar and climate. They located their farms inside courtyards where brick walls blocked cold winds. They pruned fruit trees to create espaliers on south-facing walls. With their early harvests from the hotbeds and courtyard heat islands, they beat farmers in the south of France to market with fresh fruits and vegetables. As each crop ripened, Parisian farmers piled on more horse manure and planted again, raising three to six crops a year in quick succession. Kropotkin saw in these French urban gardens a way to break the metabolic rift in which farmers removed nutrients from the soil in the countryside and shipped it far away to cities, where the resulting waste polluted the water and air while the farmers' fields grew barren.

I tried making a hotbed in a small garden plot in a courtyard in Cambridge, Massachusetts. First, I ordered seeds of Parisian "forcing" carrots from the US National Plant Germplasm bank. French urban farmers are long gone, but I relished the idea of planting seeds they had selected. Parisian farmers bred "forcing" varieties for their ability to germinate quickly under glass and to mature in cold temperatures. Most commercial farmers today use hybrid F1 seeds. Hybrid seeds do not grow true, meaning offspring will not have the same characteristics of the parent plant. So, farmers have to buy new seeds for each planting. But heirloom seeds like the forcing carrot do grow true. They allow farmers to save seeds season after season, saving money but also selecting varieties that do best in their particular conditions.

Once the carrot seeds came by post, I hauled home a pile of fresh horse manure from the Boston mounted police station. Using a market garden manual from 1911 as my guide, I dug a trench two feet deep and two feet wide, packed the manure in the hole, and covered it with four inches of topsoil. With some old bricks, I built low walls around the mound and packed the cracks with earth to make it airtight. It was late November, cold, gray, and wet. I plunged a thermometer in the center of the pile. When I returned the next day, it read 120 degrees Fahrenheit.

That hot? Surely the thermometer was broken. I shook it and put it back. Still 120 degrees. Microbes in the horse manure were creating their own little oven in the trench. After waiting a few days until the temperature had dropped to 90 degrees, I seeded lines of arugula, kale, carrots, and spinach, and then placed an old glass door over the hotbed. After that, I didn't pay much attention to it. I didn't water. I didn't weed. Now and then, I would push snow off the glass to let the sun in. The plants recycled water in the closed chamber, and no weeds could grow through two feet of horse manure. Sometime in late February, I noticed the carrots, spinach, and arugula pushing up against the glass as if clamoring to get out. I lifted the door and got out some scissors to clip the greens. We ate fresh salads from that day on. February was the new May.

The Parisian farmers Kropotkin so admired were not running boutique enterprises, nor were they remnants from the past. In 1900, farmers cultivated 2,100 acres in Paris and 300 kilometers of fruit walls. Kropotkin observed how Parisian farmers had so much composted horse manure, they sold the excess to farmers in the countryside. When forced to move, they packed up their tools and their topsoil. For urban farmers, the city was their workshop, the soil and climate their technologies. Farmers in Paris, Kropotkin insisted, could plant their farms on pavement and thrive.

At the time, professional agronomists talked a lot about soils. Soil scientists emphasized the importance of certain chemicals, especially nitrogen, for plants, and this insight became a fixation. Traders scoured the world for guano, saltpeter, and other sources of nitrogen. Farmers went into debt to buy it. Agronomists treated soils as if they were junkies addicted to a few minerals: potassium, phosphorus, and nitrogen—especially nitrogen. For decades German chemists tried to draw nitrogen, which is plentiful in the atmosphere, into beakers in their labs. They poured a lot of effort into this cause. German leaders worried that their dependence on the importation of fertilizers from abroad would be a weakness in a time of war.

It took a hundred years of experiments before the German chemist Fritz Haber devised a high-energy method using lots of heat and pressure to combine atmospheric nitrogen with hydrogen to form ammonia or nitric acid. The engineer Carl Bosch scaled up Haber's invention in a huge factory burning lots of coal. In 1909, the same year that the Oppau ammonia factory opened, one thousand people met in Budapest for the first international conference on soil science. The attendees had no idea that the "Haber-Bosch" process, an invention whose impact on human history rivals the steam engine and the splitting of the atom, would flip the ratio of nitrogen on earth from scarcity to glut. But Parisian farmers did not need this new technology. They were swimming in nitrogen, a by-product of composted horse manure in their hotbeds. Kropotkin observed: "They do not understand our talk about good and bad soils, because they make the soil themselves, and make it in such quantities as to be compelled yearly to sell some of it."

For decades landowners had been consolidating land, farms growing larger and larger to pay for expensive agricultural inputs. Kropotkin pointed to the Parisian urban gardens to argue for intensive, more

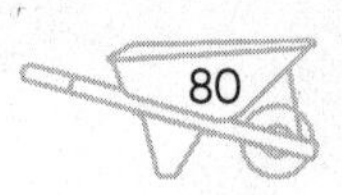

equitable smallholding. In tiny urban plots, he wrote, "everyone could grow their own food and achieve relative economic independence."

Parisian farmers proved a point Kropotkin sought to make about political economies. He rejected the prevailing social Darwinist ideas of his day. Kropotkin was the first scientist to propose that animal cooperation, not only competition, was crucial for understanding both evolution and human society. Scarcity and poverty were not "natural laws" but resulted from human institutions. Dense, crowded Paris could deliver famine; it certainly had in the past. But thanks to the innovations of the Parisian farmers, the growing megapolis served up a cornucopia. Kropotkin noted the year's harvest of one Parisian farmer: "More than 20,000 lb. of carrots; more than 20,000 lb. of onions, radishes and other vegetables sold by weight; 6,000 heads of cabbage; 3,000 of cauliflower; 5,000 baskets of tomatoes; 5,000 dozen of choice fruit; and 154,000 heads of salad; in short, a total of 250,000 lb. of vegetables." "We see," Kropotkin observed, "that it is infinitely easier to grow 200,000 pounds of food on one acre than to grow them on ten acres."

I remind myself that Kropotkin was a leading anarchist who mixed his science with his politics. Surely, he exaggerated to promote his cause? In the 1970s, an American agronomist, G. Stanhill, retrospectively calculated the efficiency of the turn-of-the-century Parisian urban farms. Stanhill found the gardens were good for the city's health and ecology, that fruit and vegetable production gave Parisians fresh food, reduced transport and storage needs, added oxygen to the air, and provided an important sink for animal waste. But what about yields? Stanhill calculated that yields were astonishing: "The Marais system of food production appears to be one of the most productive ever documented."

Kropotkin's *Fields, Factories and Workshops* was published in German in 1910 to much acclaim. But it took a war for Kropotkin's message to sink in.

Every Working Man a Garden!

IN BERLIN IN 1914, the triumphant parades sending soldiers off to war faded as the expected quick victory settled into a long conflict. The military machine consumed so much. Fats normally eaten at the dinner table were fed to guns as propellants. Artificial nitrogen engineered in the Haber-Bosch process flowed not to fields to nourish crops but to the front as explosives to kill the enemy. Before the war, the modern German diet had grown heavy on meat, sugar, alcohol, and wheat, much of it imported. The British sea blockade cut off imports, turning hunger into an effective weapon of modern warfare. During the war years, gnawing stomachs radicalized German politics, elevating tiny allotment gardens to national importance.

In the first year of the British blockade, citizens understood that war demanded sacrifice and that Germans, normally hearty eaters, would be consuming less. The question was who would eat less and how much less? People with money and people in the countryside could get food, but working urbanites began to starve. In 1915, women in Berlin stood in line all night for butter. The kaiser, the mayor, and the police chief continued to place their bets on the free market to distribute scarce goods. But free markets took an ugly, unpatriotic turn. There was nothing to stop a merchant from hoarding goods until prices climbed. When a shopkeeper refused to sell, and the line-standing women complained, he hurtled outrageous insults. "Spread shit on your bread!"

Those were fighting words. The butter riots of 1915 were followed by potato riots in 1916, which turned into bread riots by the Turnip Winter of 1917, the year the root vegetables grown for pigs became food for humans. By then, food prices in Berlin had risen to eight hundred times the prewar level, and thousands of people were dying from hunger-related diseases. At that point, the kaiser, mayor, and police chief were finally forced to take action. They tried to shape the free market into something more equitable. But they had no experience and few tools to turn competitive markets into fair distribution networks, and so the measures failed, which made rioting people even angrier.

With the state and markets defeating them, middle-class Berliners explored technologies that working-class Berliners had long practiced. Suburbanites replaced flower gardens with chicken coops and vegetable beds. The city government and the railroad handed out vacant land for cultivation. People tried to raise goats on their balconies. Hungry people gardened everywhere—in parks, on street banks and back lots—disregarding property rights.

Arbor colonists in Berlin ranked among those who ate. Gardeners grew fresh food, had stores of root vegetables and preserves, and as a cooperative bought seed potatoes in bulk for lower prices. Their small houses with low ceilings took less fuel to heat. While farmers in the countryside struggled without imported fertilizers, gardeners made their own by composting human, vegetable, and animal waste. Compared to rural farmers, shantytown gardeners enjoyed an urban heat island. When in 1915 an early frost across Germany killed potato crops, Berlin gardeners' potatoes survived just fine. As the war plowed on, the number of tiny gardens grew, stretching for thousands of acres in a ring around the working-class neighborhoods of Berlin.

Berlin and its suburbs had hundreds of acres of empty lots, but most of that land was real estate. Leasing companies treated vacant lots like merchants did butter. They hoarded and gouged. Ground rents levitated with hunger. Nurseries charged ten times more for seeds and plants than they had before the war. Hunger made fortunes.

Gardeners had been complaining about predatory pricing for decades, but during the war, the optics radically changed. Liberal politicians, journalists, and even the Berlin police began to sympathize with the hungry people who were sacrificing family members to the war effort. In the protests over food scarcity, for the first time "women of little means," as the press called them, gained voice and power. And with this new authority, gardeners won more political backing.

The minister of public works gradually eased up on regulations against sleeping in allotments. Trade unions and large factories scrambled to acquire land to set up more allotments for workers to live on and self-provision. Architects, urban planners, and politicians embraced tiny gardens as the solution to wartime hunger. The journalist Max Christian wrote that with minimal investment, "Every family can grow all of the vegetables they need for most of the year in their little garden." A landscape architect drew up plans for allotment spaces and called it the "German People's Park of the Future." Another landscape designer, Leberecht Migge, took up the banner of urban self-provisioning with a zeal delivered in exclamation points. In an enormously popular 1918 pamphlet, he proclaimed: "Every working man a garden!"

Migge, who previously designed ornamental gardens for villas, was inspired by Peter Kropotkin's attention to Parisian farmers' use of urban waste to grow fruits and vegetables. If only a third of the sixteen million German urbanites had small plots to grow food, Migge calculated, they could replace the bulk of produce formerly imported from abroad. Migge called for public ownership of land with secure long-term leases for garden cooperatives. In these independent garden communities, every man would start small by building a one-room cottage, the house growing organically with family needs. Migge designed fruit walls to block wind and retain heat. He advocated dry toilets lined with peat moss to compost human waste for fertilizer. He forecasted that after the war the gardens would continue so that self-sufficient, vegetable-eating workers could gain breathing space from the crushing forces of capitalism.

Migge's vision expanded outward. With proper planning, he pictured an edible urban landscape on a grand scale surrounding the city and spanning the nation, gardens that refreshed the air, fed people, and recycled waste. Migge's books, bought eagerly by communards from Moscow to Jerusalem, sold out.

In January 1918, four million hungry Germans, led by women, went on strike demanding bread and peace. In November 1918, Social Democrats made a secret pact forcing the kaiser to abdicate, but when the new government sued for peace, a civil war broke out. Demobilized soldiers formed into right-wing paramilitary and fought communist shock troops. Amid the political chaos, the British navy kept up the sea blockade to lay Germans so low they would accept peace on the most punishing terms. At the peace conference in Versailles, Germany lost the breadbasket of East Prussia to newly reconstituted Poland and the vegetable and fruit fields of the Rhine Valley to France. Stripped of its forced-labor plantations in Africa, Germany no longer enjoyed cheap coffee, palm oil, peanuts, and cotton.

And so peace came, but no bread. After the war, Berliners were still on their own. The Berlin city councilor August Hintze announced in February 1920, "Since the country is currently unable to meet its food obligations to the urban population, urbanites will have to help themselves and cultivate their own gardens to provide the necessary food." The demand for allotments grew daily. Several hundred thousand refugees, displaced by Germany's shrunken borders, poured into Berlin. Inflation wiped out wages and life savings.

One of the first acts the new German Weimar Republic president Friedrich Ebert passed was a law guaranteeing the right to garden. The July 1919 Small Gardens Act delivered protections gardeners had long sought: rent control, long-term leases, and security from arbitrary evictions. The Weimar minister of welfare mandated that cities across Germany create allotments and settlement offices that would make public land available for garden colonies. Three hundred thousand people joined a newly founded union of German small garden associations. The Berlin government earmarked funds to design and build allotments with housing for workers. With public money in the mix, more professionals jumped on the garden train. The modernist architect Bruno Taut joined Migge to draw up plans for affordable garden housing in Berlin.

When Migge wrote *Everyman Self-Sufficient!*, Berlin gardeners already cultivated more than sixty thousand allotments across the city. After 1920, when Berlin absorbed its suburbs, the city contained over fourteen thousand acres of urban gardens, 7 percent of the city territory. Berlin gardeners flew green and white flags over their huts, a symbol of their allegiance, not to the nation or this or that political party, but to their cooperative cause. Migge acknowledged the existence of working-class gardeners and he sought to support them, but he also borrowed heavily from their practices. He laid out plans for small, modular housing, composting, intensive gardens, wall orchards, cooperative management, and mixed economies as if they were novel inventions.

Great men are said to produce great ideas that materialize as if from nothing other than synapses firing in their great brains. By appropriating the inventions of working-class gardeners, Migge left behind an enfeebled historical imagination of the range of human possibility. He reconfigured real, spontaneous, garden communities as if they were a utopian vision for a better future. But there was nothing utopian (meaning impossible) about them. They already existed.

It is hard to express how swiftly and powerfully the politics of simple urban gardeners took center stage during the hungry years after the war. As allotments became part of the city bureaucracy, Berliners asked that the staff include working-class gardeners in the administration "so that their painful experiences can be included in the [administrative] process." Thousands of Berliners attended rallies, charged with the possibility that the new republic might deliver what they sought—stability, protection, and support so that "even the simplest worker has the opportunity to live in a home with garden access." Songs, poems, and speeches repeated this straightforward request: "*Gebt ihnen Gärten!*/ Give them Gardens!" A city official complained: "Some daredevils even want to give every urban worker the possibility of an allotment garden." Experience had taught working-class gardeners that they could organize themselves, but to persist they needed the state to shield them from the market. With state support, however, came control.

Charlotte Tessen and her husband moved from one arbor colony to another in Berlin during the 1920s until finally they rented an allotment. The couple landed on a former field divided into lots in Wilhelmsruh. They paid 18 marks a year for six hundred square meters. They had no electricity and no water. They had a small handcart to carry broken bricks from a rubble yard to build a foundation. They erected walls for a tiny hut with wooden battens and tar paper. They found a driller to dig a well. A neighbor, a bricklayer, helped them construct a chimney. In the summer ants nested in the walls, and they dug them out. Tessen's mother came to live with them, and they built an annex for her. They set up a stall for chickens and rabbits. A few

*Mother with her child in a residential arbor colony.
Wedding-Berlin, 1929.*

years later Tessen gave birth to her daughter in their growing home. She remembered fondly her life in the green shantytown, where children ran about freely between apple trees and flowers.

City leaders set out to improve on such informal settlements seen as dirty and dangerous "gypsy camps." In 1920, the city of Berlin earmarked 3 million marks to build a garden residential community for the very poorest Berliners living in tenements in an industrial region known as "Red Wedding" for the number of communists living there. The project took shape on a former heath, the Jungfernheide, which, after a poorly conceived drainage project, had become a vast sand dune. Initially, the plans for the colony of two hundred houses reflected designs of working-class arbor colonies, like Tessen's, that were already established on the site. The first proposal called for the conversion of the existing summer cottages into permanent living quarters and the construction of more small, wooden houses with turf insulation. The houses would be connected with dirt paths, gas, and water, but no sewers. Each house would have a garden of a tenth of an acre. Tenants would have leases up to thirty years. The proposed houses were cheap, quick to build, tight, comfortable, and approved by members of the allotment garden committee who, as a city council put it, "were the experts after all."

But soon professional architects and landscape designers got involved and shouldered out the gardeners and their designs. The architects did not understand the insulating properties of turf. They complained the wooden walls filled "with dirt" would rot. They proposed instead to build cottages from cinder blocks, a newly manufactured material made from coal ash that insulated poorly and deteriorated quickly. The city architect said the cabins were too small for a family, and he doubled their size. Rather than footpaths, he advised wide streets paved with gravel. A landscape architect designed a park around the development with playing fields, wading pools, a running track, and garden allotments. As the costs went up, the councilors shortened the leases from thirty to six years and raised the rents. Once professionals got involved, the housing, initially designated to shelter "the poorest of the poor," was priced entirely out of their reach. By the time the project went up, only the most affluent workers could afford the rent. The public funds allocated to build the settlement supported people earning good wages, not the unhoused, not those scraping by.

To make way for the new development, city workers cleared several existing arbor colonies. Amid a severe housing and food crisis, they tore down houses and gardens of several hundred people to house and feed several hundred other people. "Overnight," an observer wrote, "everything changes from splendor, and the land is churned up for a world of stone." The development swept away the self-organized community. New people took over. Officials selected families they thought best for the project and laid out rules for planting, behavior, and maintenance.

The Small Gardens Act was designed to protect working-class gardens, but across the city, officials sanctioned developments that pushed out arbor colonies organized and managed by workers, while gradually taking political control over Kleingärten (small garden) allotments enrolled in the Reich Kleingärten Association. "Small gardens" referred to the "small people" (*kleine Leute*) who inhabited them, the scent of paternalism wafting from the name. Gardeners in Kleingärten felt watched. They cautioned each other against saying too much because "We are constantly monitored by the Reich [Small Garden] Association." With such oversight, people lost interest in attending meetings.

Kleingärten had official sanction and were organized, while informal arbor colonies, populated by the most impoverished Berliners, were seen as undesirable and were under assault from development pressures. As gardeners scrambled to save their arbor colonies, they wondered how, after having finally won the right to garden, did it go so wrong?

The Weimar ruling party, the Social Democrats, stacked ideas from the left and right into a strange political sandwich. One ingredient they added to the mix came from the eugenic menu of the day. In Germany, the left-wing tradition of worker's self-realization through arbor colonies existed alongside right-wing dreams of purifying the German body politic in garden cities that were to serve as grounds for selection (*Auslese*) of the best and elimination (*Ausmerze*) of those determined to serve no useful purpose. Urban planners sketched maps grading working classes according to utility, recommending that the shiftless poor, who were a "constant burden to the state," be removed entirely. Darwinian selection was on many people's minds. During the Weimar

Republic, Gustav Böss, Berlin's Social Democratic mayor, observed: "Everyone takes the breeding of horses for granted. That man, too, must be bred should be more obvious."

Germany led Europe in creating the leading institutions for social welfare. But they designated public funds to finance middle-class relief, while poor people were left to their own devices. Eugenic ideas slipped from books and conversation and spilled out across the landscape to shape the foundations of cities today.

Germans, however, were not leaders in the field of eugenics. At the time, people who wanted to know more about this growing discipline of pseudoscience went to the world leader on that topic, the United States.

THE THIRD TURN

Washington, DC

A Quarter Acre and a Hog

IN 2015, THE CITY OF WASHINGTON, DC, sponsored a design competition. Architects were tasked with drawing plans for green housing in Deanwood, a Black neighborhood that, like much of the rest of the city east of the Anacostia River, suffered from high rates of chronic disease, crime, unemployment, and poverty. The architects' challenge was to design ten affordable, sustainable houses. The point of the contest was to show that green design was not just for the wealthy.

The winning team presented plans for a row of small houses surrounded by gardens. As access to clean water will be a problem in the future because of climate change and population pressure, the houses had composting toilets and filtered rainwater harvested from the roof. A third of the lot space went to growing fruits and vegetables for residents. The houses opened onto a common courtyard, which the architects hoped would enable the community to "feel safe and connected to each other." The designers emphasized that the goal was not just affordable housing, but affordable living. They hoped that living partially off-grid would emancipate Deanwood residents from the chronic problems of making rent and obtaining healthy food.

After the awards ceremony, the design team took photos, packed up their models, and went home. That was it. Financing obstacles meant the green housing plans were stillborn.

No one at the event noticed that the award-winning design pointed not to the future but to the past. Green architecture is said to have started in the 1970s during the global energy crisis, when white professional architects and designers drew up plans for housing that recycled, composted, and generated fresh water and food. But in the 1920s, Black working-class men and women were already forging what today would be called "green" and "sustainable" communities in Washington, DC.

Very helpful archivists in several DC repositories helped me locate oral histories of DC residents taken from 1975 to 2020. Alice Tate was born in 1929 in Marshall Heights, east of the Anacostia River where the design competition took place. Her father had a large urban lot on former "rich farmland." Nearly a century later, in 2020, an interviewer asked her about it.

"He had a garden. He had pigs, hogs, chickens. Apple orchard. Grape orchard. Lady slippers. Dogwood trees. Pink and white. People can't imagine how beautiful it was. . . . It wasn't just a small apple orchard. He had about ten to fifteen apple trees, and he had grapes all over the place. He had a peach tree and he had gooseberries. I mean, he had corn and everything. He had beautiful flowers. . . . Mother canned. Green beans, corn, tomatoes, peas."

When asked if her grandmother sold her canned goods, Alice's daughter, Loretta, chuckled. "No, she'd give them away."

Alice concurred, "She gave half away. And we had homemade sausages. Father did not sell sausages. My parents made wine."

Loretta: "Good wine."

"Was there anything that he did on the farm to make a profit?"

"No."

"So," the young interviewer tried to understand, "He just wanted to be able to have these things . . . ?"

Loretta: "So they could prosper."

To prosper as a Black household in DC in the first decades of the twentieth century was no mean feat. Since the Civil War, Black migrants from the South had been moving to the city in a steady stream, and white Washingtonians had been working to cordon them off in certain areas. As DC grew and spread, Black working people were confined to older neighborhoods of small houses on alleys.

For decades, white elites fretted over those alley dwellings, a "men-

Woman in the doorway to her alley house. Washington, DC, 1935.

ace," a "running sore and disgrace to the Capital." Much like their German counterparts in Berlin, elites in DC went on ghetto tours, dignitaries in wool suits and silk scarves striding down alleys, the strong scent of eau de cologne mixing with the punch of ammonia from wooden box toilets. Inside a home, a senator nervously gathered his coat skirts, touched nothing, and hurried to leave. The visitors were appalled to find a mother and four children sleeping in one bed together. One-room cabins might be OK for white settlers in the West but were evidence of moral degeneracy among the urban poor.

According to experts, "the Negro" was poorly rooted in the landscape. "The Negro" spread blight, disease, and social problems. Zora Neal Hurston observed that, like a taxidermy specimen, "the Negro" was all frame, wire, and straw, hollow inside and meaningless, but useful to pull out when occasion required it. Unfortunately, this empty form played an outsized role in shaping the American urban landscape.

City officials condemned alley houses and pulled down several hundred a year after the turn of the century. But as they leveled houses, few new ones were built in their place. Black alley dwellers had little choice but to move to another alley, subletting now just part of a room, crowding in even more densely. For landlords, the alley houses gave "a large return on the investment," and so the problem did not go away. Meanwhile, the number of Black residents in DC grew to 30 percent of the city population by 1920, making for the largest Black urban community in the United States. This demographic fact bothered many white Washingtonians, and they tried to correct it.

President Woodrow Wilson curtailed Black people's access to skilled and professional jobs in the federal government. Real estate developers drew up covenants to bar Black residents from DC neighborhoods. Police harassed Black Washingtonians with anti-vagrancy laws. In 1919, when a mob of thousands of white men marauded in Black neighborhoods, killing six people, police stood by. Vigilante violence, abuse, and housing discrimination forced Black people into ever-tighter corners of the city. Norman Dale recounted to an interviewer in 1975 about his youth in the 1920s and 1930s: "Places you could not go . . . The average black person in those days . . . it was so ground into him that you're just nobody."

We usually think of the Great Migration of six million Americans from the rural South to urban areas as having severed Black people from the soil and estranged them from traditional foodways. This is certainly true, but other histories run alongside it. In these histories, people reimagined themselves as they wanted to be. They sought a place in the sun where they built, dug in the soil, and found a cooperative liberation.

As the noose of segregation tightened, Black Washingtonians began to move, slowly at first and then in greater numbers, across the Anacostia River to neighborhoods like Deanwood and Marshall Heights that had parcels of land without restrictive racial covenants. The location was remote, semirural with little in the way of infrastructure or services, and it contained some unpleasant features. Along the Anacostia River, a large abattoir processed meat. When the plant boiled offal to make grease and fertilizer, an "unbearable" odor floated into the air. Across from the slaughterhouse, the city's coal-fired power plant pumped out a black soot and dumped cinders into the river. Erosion and dredging reduced the river's depth from forty to four feet. The newly shallow river flooded more frequently, creating mosquito-rich mud flats that triggered deadly malarial outbreaks.

But once you got past the river, areas East of the River had some advantages. Hills carpeted with towering oak, beach, elm, and chestnut offered shade, fresh breezes, and gorgeous views of the capitol. Springs bubbled from the hillsides, feeding streams where fish swam. And East of the River had history for Black people. Frederick Douglass's Cedar Hill estate was a magnet for Black intellectual life. In 1897, Civil War refugees, called "contraband" and expelled from the city center, established a community called Barry Farm, where they planted gardens and orchards, raised small animals, and built churches and a school for Black children. Just to the north, in Deanwood, the Congee African Methodist Episcopal Zion Church was founded in 1884. And in 1909, the brilliant, indefatigable Nannie Helen Burroughs raised money from Black Baptist churches to establish in Deanwood the National Training School for Women and Girls, the first vocational school in the country for Black women.

In the 1910s, classified ads targeted Black buyers to purchase lots East of the River in Deanwood, Burrville, Lincoln Heights, and Marshall Heights. A streetcar terminated in Deanwood, which made it most desirable. Residents founded a grade school for Black children in 1912, building and furnishing the school with money from bake sales. Recognizing that Black parents worked long hours, administrators created an unusually comprehensive program that included a kindergarten, a day school, a night school, and a school garden, where every child had a plot of their own. Deanwood had the first and only amusement park in DC, Black-owned Suburban Gardens.

The neighborhoods grew in patchwork fashion, a few houses set down among tracts of empty lots. Working with neighbors and teams of horses, people dug foundations and moved cottages from other parts of the city. People fashioned houses with whatever they found—egg crates, beaver boards, and recycled planks. One house was made of old telephone booths. Ministers, tailors, and waiters got good at carpentry, bricklaying, and cement finishing. As members of an extended family arrived, a house would grow, rooms rambling this way and that. The small boxy structures were surrounded by "highly cultivated" gardens.

In the thirties, two white real estate developers, Howard Gott and Joseph Tepper, bought up many lots in the area. They offered credit to Black families. Interest rates were high. Gott would walk up and down streets on Sundays with a little book, collecting mortgage payments. A few people remembered having trouble shaking Gott off once they had paid in full because he refused to close out the loan. Miss a few payments and a family lost all the money they had paid down. Gott foreclosed on a lot of loans, but people still risked it. "There has been an unshakable conviction among Black people," the writer Natalie Baszile observes, "that true liberation requires landownership."

When an oral historian asked Norman Dale what sort of jobs they had in the 1920s and 1930s, he replied: "There weren't too many of them who had jobs. I would say there was a very high unemployment rate." Alexander Davis remembered: "The only way you could survive was by doing a little farming."

Real estate records show that most buyers put money down not on one lot but on two to six lots. Vincent Bunch's father bought six lots in Deanwood. The family constructed a small house in the center of the property and made gardens out of the rest. "We planted everything. . . . Corn, sweet potatoes, peanuts, greens, squash, peppers, and tomatoes. Seven peach trees, two apple trees and one plum tree that didn't bear too much, and a cherry tree. Come to think about it, we were pretty self-sufficient."

Willie Hardy remembered, "We raised hogs and chickens and grew most of the food we ate."

Ethel Green recounted: "We had an acre down there, raised everything, beans and chickens. Three-hundred chickens, fruits and berries of all kinds, flowers, and we sold them in market. . . . Mother had stands in the old center market, sold honey, flowers, fruits, and vegetables. Father lost the first house. Mother bought the second one."

Taylor remembered his parents' household: "They got along very well. . . . We had two cows. We used to sell quite a bit of milk around the neighborhood. We also had gardens on both sides [of the house]. . . . Big trees, gorgeous trees, chestnut trees. Plenty of fruit."

It might sound insignificant—growing vegetables on the side of the house, eating them and selling off the extra—but there was serious money to be made. At the time, fresh fruits and vegetables were priced as luxury foods. In 1917, the US Department of Agriculture calcu-

Villagelike Marshall Heights with fruit trees and animal pens, March 27, 1949.

lated that a small Washington backyard garden (272 square feet) would require $4.82 in investment, about fifty-eight hours of labor over a season, and generate $1,306 in fruits and vegetables, more than double what most people in the neighborhood could make in a year of wages.

A few small animals added to the pot. Clarence Turner in 2020 remembered: "We had chickens in our yard. [My father] could get about 27 dozen eggs a day. He would sell them." People also traded in produce. "You could always tell when Lawyer Moss had a case," Pierre Taylor reminisced. "You'd see him coming home from the police court with a goose or a chicken or a duck under his arm." Harold Thompson moved to Deanwood in 1925. He recalled plenty of "truck farms": "Well when you say truck farms, I don't mean the commercial type. Most of them were just families, with their own farms, and using what they needed for the most part."

As white-owned commercial farms were closing around DC, residential housing taking their place, Black homeowners filled the gap, growing many thousands of pounds of food on city lots. The USDA census shows that although the acreage of farmland in the city declined by half, the number of pigs and chickens, the usual livestock of Black home gardeners, doubled from 1920 to 1925, even though nuisance laws by that time had banned hogs and chickens from the city.

Unlike their white neighbors who consumed a diet heavy in wheat, meat, and sugar, Black Washingtonians ate from an astonishingly rich and varied menu. In the process of extending a metro line, archeologists found seeds of grapes, peaches, plums, cherries, squashes, pumpkins, watermelons, beans, greens, and corn. They pulled out the bones of rabbits, squabs, possums, chickens, and hogs.

Of course, as Alexander Davis pointed out, former sharecroppers, people like his parents, were "tired of farming." As tenant farmers in the South, "They'd work all the summer and at the end of the year the owners would [cheat them] and then they have to suffer through the winter." But gardening is not farming. Anthropologist Ashanté Reese points out that the same agrarian tools that Black Southerners associated with subjugation, residents East of the River used to develop individual and community self-sufficiency. Former sharecroppers understood that access to land offered security and a salve for hard times. For them, freedom was not inscribed in law. Freedom was spatial.

DC was the only city in the country that had no democratic governance. Congressionally appointed commissioners ran the city with oversight from senators and congressmen. The district commissioners and real estate developers who built infrastructure for white and middle-class Black neighborhoods did nothing for poor Black neighborhoods. Property owners were on their own. Willie Hardy remembered Marshall Heights as "wilderness": "There were no streets, no lights, no water." Mildred Tompkins recalled the mud. Children walked a mile and a half to school and adults went on foot to work: "about four miles for my man," a woman told a reporter.

Years passed, then decades, and Black residents East of the River remained without much in the way of public infrastructure. In the 1920s, the commissioners who managed DC were increasingly Southern segregationists concerned about Black "invasions" of white areas. The congressional supervisors skimped on paying for roads, fire stations, public transportation, garbage collection, libraries, and schools. As one Black paper put it: "The District government cared as much for a negro as a rat terrier does for a mouse." Black residents East of the River had to fight for every paving stone, every bus, streetlight, and water main, and that struggle went on for decades. But in the meantime, people East of the River put their marginal status to use.

Norman Dale: "We didn't have garbage trucks in those days. You burned the trash and the garbage, if you had pigs, you gave it to the pigs and if you didn't, you made compost out of it." Assuming that the interviewer did not know what compost was, he explained: "Compost is fertilizer for your garden. You would set aside a corner of the garden for a compost pile and the more garbage or any kind of food, you would put on there, pile leaves on it and other cuttings and trimmings and in two or three years you had some real fertilizer to go on your garden." Roving pigs cleaned up streets. Chickens pecked through food scraps and kept insects at bay.

Even the waste of backyard privies was upcycled. There is a common assumption that once flush toilets appeared, urbanites quickly adopted the convenience, but in DC, the volume of night soil did not fall, but rose decade after decade until the late 1940s. Rather than flushing human waste away, many Washingtonians prized it. An agronomist for the USDA wrote: "If I wanted to redeem a piece of ground that is considered worn entirely out and make a garden of it, give me that which comes from cleaning privies."

In Black neighborhoods East of the River, more than a third of houses had outdoor privies. Leaking outdoor box toilets caused illness in other parts of DC, but cholera and typhoid did not haunt the settlements across the Anacostia River. Barrel men cleared backyard privies and brought the waste to a dump run by the Washington Fertilizer Company. There Black sanitation workers fermented the night soil into compost and sold it to gardeners and farmers. Sanitation workers emptying garbage cans in central DC pulled out bread, which they sold to gardeners to feed pigs. In 1921, a garbage collector, paid $3.50 a day, could triple his income by selling bread scraps. Hundreds of hogs roamed the Washington Fertilizer Company's dump. The hogs ate food scraps, reportedly leaving the territory clean. Garbage was so important to the urban hog economy that people fought over it. When the Washington Fertilizer Company complained about lone prospectors pillaging garbage cans, the district commissioners passed an ordinance barring garbage collection to any but the city contractor.

Urbanites ate and relieved themselves. Night soil collectors gathered up the waste and brought it to the dump, where they turned it into fertilizers, which gardeners used to grow produce that urbanites ate. This closed, organic cycle worked pretty well, except at times when district commissioners stinted on appropriations to pay for safe hauling. Bacteria in the right place are incredible producers of wealth and well-being. Bacteria in the wrong place leads to tragedy. DC residents whose houses connected to the sewer system flushed 90 million gallons a year of sewage into local waterways. Even after the city built a larger waste treatment plant in the 1930s, overflow sewage poured into the Anacostia River for all of the twentieth century until 2018.

Reacting to the waterborne epidemics that were often caused by the new flush toilets contaminating water sources, Louis Pasteur and his followers declared war on bacteria and other microbes. Public health officials believed the best response to "germs" was total eradication. Across the United States in the 1920s, developers ran sewer lines, paved green spaces, filled in marshes, leveled hills, and channeled springs, streams, and creeks into underground culverts.

But scientists are realizing that microbes living on and in the human body are the major architects of human life. They determine how a person feels and how well a body functions. We owe our lives to the society of microbes dining, reproducing, fighting, and collaborating inside human bodies. As city dwellers isolated themselves in territories of asphalt and stone, while spraying streets and homes with bug-and-bacteria-killing chemicals, whole species of microbes that had for millennia lived alongside and inside humans disappeared. Scientists don't yet fully understand how, but the loss of diversity in the human microbiome correlates with chronic diseases such as diabetes, obesity, cardiovascular disease, and certain cancers. Now scientists recommend that for good health, people, especially children, should spend lots of time outdoors, dig in the earth, and have contact with animals.

In Black neighborhoods East of the River, that is just what people were doing because of city leaders' neglect. Streets remained unpaved. There were no sewer lines or garbage pickup. Chickens, dogs, cats, and hogs roamed about. Streams flowed above ground over streets of packed earth and threaded through reedy stands to pool in swimming holes that drew children, insects, birds, and animals. The creeks filtered rainwater that percolated to underground aquifers. Residents drew water either from wells or what they called a "system," barrels that captured rainwater from the roof and filtered it through gravel. Trees reached down to tap into the groundwater. Without pavement, tree roots were free to stretch out and do what they do. Elders recalled the groves of fruit trees. They spoke of woods with century-old oaks and chestnuts. Vincent Bunch described the pecan tree his mother brought up from South Carolina. Permanence was a luxury few Black Americans could afford. Transplanting a tree or flower bulbs helped migrants connect the old place with the new and build a sense of home. Bunch said the tree never produced much, but he was happy it grew, a living tribute to his mother.

Most everyone who participated in oral interviews recalled not the contours of their houses but what was outside: the hillside spring, fruit trees, hardwoods, frogs, squirrels, rats, possums, neighbors, and friends. Their descriptions demonstrated how people, as they constructed homes and gardens for themselves, also left room for other species to prosper.

Slowly, communities East of the River pieced together alternative infrastructure and economies. The Deanwood Civic Association hired Bernie Chapman to run a community bus service to drive people back and forth to the streetcar. The Civic Association wrote petitions to the district commissioners to lobby for schools, to hire Black residents for jobs in schools, and for fireproofing and repairs after the shoddily built schools went up. Residents formed what they called "family" churches, little one-room meeting houses where an extended family gathered to pray. On Sundays, parishioners passed the plate. Women baked fruit into pies for bake sales and cooked up garden vegetables into dishes for church dinners. With the money they earned, they took kids on trips and handed out small loans to those who needed it. In the winter, churches distributed piglets to parishioners to raise. In the fall, the community would gather to slaughter the fattened hogs. Dale remembered: "They would do a co-op type of thing. . . . Killing time, they would go from one family to another, trading off and trading help. They would pool their help to kill and to cure, and all this sort of thing."

Down the road at the commercial slaughterhouse, meat packers toiled in a race to dismember as many animals a day as possible, while in nearby Black neighborhoods, a hog slaughter was a celebration. As Loretta Tate recalled, "The whole community would come together and they would put the pig on a spit."

A church picnic was a gathering of epic proportions if you look at it from the ground up. A gardener would pull up root vegetables and shake out the dirt. She would cook the potatoes and carrots and bring them down the street to a church potluck. Others would come carrying their dishes in baskets. The picnickers would greet each other with hugs, kisses, and handshakes. A few men would hand out roasted corn and cuts of pork sizzled over a fire. Others would pass along freshly baked rolls and slices of peach and cherry pie. Through it all, children would run, girls combing out each other's hair, boys wrestling. Adults would catch up a toddler who fell to the ground. Over millennia, humans invented practices that enabled them to absorb microbes from their surroundings and exchange them—hand in earth, hand to hand, hand to mouth—back and forth, a wealth of microbial life passing between them. Scientists now understand that people who share microbes share much else in common, including a greater social intimacy. That is how people make kin. In memory after memory, residents East of the River spoke of the strength of their community and how much it gave them. Their communities were biological as well as social; the bonds they built cultured a rich brew of connections and collectives that saw people through tough times.

Shantytown in the Woods

IN 1929, THE STOCK MARKET CRASHED, and banks shut their doors. As the depression deepened, shantytowns, called "Hoovervilles," spread across the country. The image of "Hoovervilles" helped propel Franklin D. Roosevelt into the presidency. Roosevelt's administration laid out a host of new programs designed to use public money to stimulate the economy and bail out bankrupt businesses and farms. New Deal reformers conjectured that if struggling citizens had just a bit of land and some communal organization, they could self-provision through economic downturns. In 1933, the Roosevelt administration set up the Division of Subsistence Homesteads and allocated $25 million to help Americans buy land, build affordable homes, and "raise a part or all of their food." California Governor Upton Sinclair suggested that federal funds should support already existing shantytowns by building infrastructure for water and sanitation. But, as with state-sponsored housing projects in Berlin, New Deal officials dismissed working-class design and planning. Instead, architects working for the homestead division set out to fabricate new subsistence settlements. They drew up plans that grew more elaborate each rendition. Sociologists came up with applications to vet homesteaders for their race, intelligence, and health in line with eugenic science of the day. In the end, the Division of Subsistence Homesteads built thirty-four segregated communities at prices too expensive for a poor person to afford. The program did little to help the majority of Americans who were struggling to find food and shelter.

But just a few miles away from the homestead division's offices on Constitution Avenue, Black residents in neighborhoods East of the River had already created cooperative communities of people who raised all or part of their food while living in affordable homes.

Like other New Deal projects, the Division of Subsistence Homesteads favored white, middle-class Americans and business interests while most often leaving Black Americans to fend for themselves. Federally insured housing loans went to white Americans because realtors determined that loans for houses in Black neighborhoods were too risky. Thousands of Black migrants ejected from farms in the South arrived in Washington to find few housing options except for rooms in crowded alleys. In 1934, alarmed about the dangerous congestion, district commissioners created the Alley Dwelling Authority and vowed to eliminate all alley housing in ten years. First Lady Eleanor Roosevelt got involved. She spotlighted the human misery in the alleys and campaigned to pull houses down. More people headed East of the River.

There, Black subsistence homesteaders cut streets through woods and built housing in an even more haphazard fashion than before. In a nod to infrastructure, the city installed hand-pumped wells on a few street corners so that residents in emerging neighborhoods like Marshall Heights could have clean water. The city extended no other services.

And so people got by the best they could, pitching in to help each other out. When a house caught fire, it would take forever for city firefighters to arrive from across the river, but neighbors would take immediate action. Clarence Turner spoke about a fire that leveled a house with many children, "We built a place big enough for them to move in that night. . . . Built the walls and put a roof on it in one day. . . . And after that, they built a house on the property. I'm telling you, people helped each other out then."

Nearly every interviewee from 1975 to 2020 spoke, as Turner did, of a neighborliness that was both social and commercial. They emphasized that people knew everyone else in the community and they helped each other out and hired within the neighborhood to do jobs. Children knew to carry water and groceries to the elderly. They checked on a neighbor who fell sick or had a baby. They watched out for each other's children. They gave beds to newcomers who arrived from the South in the chain of migration.

When in 1935 a white man sought to open a store in Marshall Heights, residents fought it, even though they had no grocery. Reverend James White told a reporter, "We plan to have our own cooperative stores out here. And we will have our own boys and girls working in them."

And that's how it went, more or less. In the "bottoms," where a swamp formed on the mud street in rainy weather, people set up shops, selling candy and dry goods from their kitchens. On the hill, a juke joint served up alcohol, chicken, and music. Women sewed and catered from their homes, and hucksters pushed carts up and down the streets pedaling fresh produce, fish, and ice. Midwives delivered babies that were placed in dresser drawer cribs until the doctor came to check them out. The profits these informal businesses made cycled back into the communities.

In the 1930s, neighborly mutual aid formalized into cooperative movements that were sweeping the country. In DC, activists created the Central Union Cooperative and the Young Negroes' Cooperative League to coordinate self-help initiatives on the neighborhood level. In Marshall Heights, fifty men formed a cooperative to deconstruct condemned structures and use the materials to build new houses. Women created the Willing Workers cooperative and sewed sheets, dresses, and other household goods. In nearby Deanwood, Mammie Montgomery managed a seven-acre community garden on vacant land. The educator and civil rights activist Nannie Helen Burroughs founded Cooperative Industries, which provided medical services, a soup kitchen, and workshops that gave jobs to unemployed domestic workers. The women sewed, canned, baked, repaired shoes, and made mattresses. In 1936, after Cooperative Industries won $19,000 in federal funding, the organization expanded, starting a farm nearby in Maryland in collaboration with the Tuskegee Institute. The cooperative farm ran a roadside market and sold fresh produce, meat, poultry, and dairy at the co-op's grocery in Deanwood. Revenue from sales went into a fund for education and relief.

Deanwood and Marshall Heights were not unique. Black Americans created cooperative garden neighborhoods on urban peripheries across the country, places such as Orange Mound in Memphis, Eight Mile Wyoming in Detroit, and Chagrin Falls in Cleveland. Italians, Poles, Irish, and other new Americans also dug into vacant lots, roof and backyard gardens in cities across the Midwest and eastern seaboard. Asian Americans and Mexican Americans circled West Coast cities with lush market gardens.

A sharing economy makes sense in a garden terrain. Plants are generous. With care and good soil, plants grow and donate fruits in excess. Gardeners find it easy to act in the fashion of plants because in season they have plenty of fresh produce to go around. Alice Chandler told an interviewer: "Your neighbor may have tomatoes and squash in their garden, and you may have cucumbers in yours. Depending on how bountiful each one was, they would trade off." Chandler recounted how neighbors would go fishing. "They would bring back enough for friends in the neighborhood. That often meant a Saturday evening fish fry at home."

Spending time in Marshall Heights, my friend Kenda Mutongi and I stepped outside and heard a voice calling out "Hallo!" We fell to talking to Wanda. She had been heading over to her brother's house, but on meeting us, she changed her plans and gave us a tour. We walked up one hill, then down another, past older houses that extended like an accordion, each addition to the house pulling the building to the back of the lot.

Wanda, born in the late 1960s, remembered all her former neighbors. She named the owners of the standing houses and the ones pulled down.

"It was fun growing up here. We kids ran wild."

"Right here," Wanda pointed to a row of new townhouses, "was all forest."

Wanda had nothing good to say about the new housing going up. In the past decade developers had been buying up low-priced housing to build expensive condos. Neighbors were suspicious of people like me, white outsiders. Washington is a small city, but the quadrants, where people group up by class and race, are worlds unto themselves. I had lived in DC for two decades, but I had crossed the Anacostia River only a handful of times.

We stopped in front of the house of Wanda's grandmother, where Wanda lives now.

"We had a peach tree right there in front of the window, then a plum, pear, and apple. On that side were blueberry trees." The fruit trees and blueberry bushes no longer existed. Wanda came of age in a later generation when tending plants other than turf grass marked a family as poor.

Wanda pointed to a small handkerchief of a garden in front of the house. "My grandmother grew vegetables here, enough for us and our neighbors."

"In this little spot here?"

"That's all the space she needed. The soil here is very good. Greens still come up that I didn't plant!"

Wanda, like others we met, remembered the exact location of long-gone orchards that neighbors treated like a commons. "Anyone could pick." In the 1940s, a white sociologist asked residents of Eight Mile Wyoming, a Black garden community in Detroit, if they had trouble with theft from their plots. A woman replied, "Trouble? No, we don't call it that. You mean, if a man goes by a nice cornfield and he sees nice ripe ears with yellow tassels blowing in the wind—does he just take a few? Why, sure he does! We just plant a little more than we need each year to take care of that. We never have no trouble at all. If we run low, we just get a few off somebody else's. We'll know that. We don't care. We're friends out here!"

Where does generosity and sharing edge into politics? Almost no one in dozens of oral histories said they were politically active. "We knew nothing about politics," George Trivers insisted. DC residents, ruled by congressional fiat, had no elections, no reason to join a political party or attend a rally. Black and white residents formed citizens' (white) or civic (Black) associations to lobby district commissioners, but Black residents were barred from official meetings in private homes and whites-only restaurants and hotels.

Yet, residents' sharing and cooperation resonated with contemporary political voices in the Black community. The preamble of the 1936 constitution of Nannie Helen Burrough's Cooperative Industries stated that the organization aimed to redistribute wealth through mutual self-help. W. E. B. Du Bois promoted cooperatives as a way for Black Americans to resist the structural racism built into American markets. One goal of the Universal Negro Improvement Association was self-sufficiency through food production and Black-owned business.

George Washington Carver at the Tuskegee Institute called the black-eyed pea a "mortgage lifter" because it restored soils, drawing nitrogen from the air to feed other plants as it grew. At a time when agronomists promoted the adoption of heavy machinery and the use of cheap, new artificial fertilizers made from the Haber-Bosch method, Carver, like cottagers in nineteenth-century England, advocated working soils with simple shovels and fertilizing with what gardeners had on hand: fallen leaves, swamp muck, animal manure, and human waste. Commercial fertilizers, he showed in his research, "will stimulate and for a while produce good results, but by and by a collapse will come, as the soil will be reduced to practically clay and sand."

Carver showed astonishing improvements in fertility and water-holding capacity when he applied his locally scavenged compost. "A walk through the woods," he coaxed farmers who went into debt buying imported guano, is a stroll through a "natural fertilizer factory." Instead of drawing on fossil-fuel-intensive chemical fertilizer, Carver understood that relationships between the animal, mineral, and vegetable kingdoms allowed cultivators to thrive without draining resources and degrading ecologies elsewhere.

Black communities drew on a long tradition of customary possession and common right. The historian Monica White describes such "everyday utopias" as place-based alternative practices and experiments in everyday living. Residents East of the River left few political treatises, but their actions—building, producing, planting, collaborating, and sharing—sounded out an ideology of liberation, "a realization of democracy in industry" (Du Bois's words). This sustainable self-sufficiency promoted a quiet flourishing.

Not that there was a Shangri-la kind of universal solidarity. Social and racial fault lines ran through Black communities East of the River like anywhere else. Some parents told their children not to play with the poor kids who lived in the bottoms. Some people believed that the lighter a person's skin color, the more desirable they were. Middle-class Black families who lived nearby in Capital View, a neighborhood of "spic and span" houses, barred Marshall Heights neighbors from joining their civic association. Homeowners who had spent $4,000 on their homes feared that an alliance with Marshall Heights neighbors (whose houses cost "upwards of $50") would deflate their property values.

And there were plenty of outside critics too. Eleanor Roosevelt showed up in Marshall Heights in 1935 wanting to know what happened to the people her campaigns had displaced from the alley houses in central DC. The First Lady, whose toilet burped untreated feces into the Potomac River, lived in DC's Penn Quarter, which was clean at the expense of other city neighborhoods. She was appalled to find makeshift cottages with outhouses and compost piles. "These people in Marshall Heights," she pronounced, "are in a worse condition than they were before they moved."

It is too bad that the First Lady did not ask anyone who lived in Marshall Heights what they thought. Florence Collins, reporting for the *Afro-American*, did. Inside the modest houses, she found cozy interiors, decorated with various samples of wallpaper in all colors and patterns. Despite the lack of running water, Collins found the homes to be exceptionally clean. The roof might leak, another journalist in 1935 recounted, but "There appears to be a real love of country here."

Nor did the First Lady grasp the importance of a title to property. Homeownership was a significant sign of financial stability. Americans invested most of their savings in housing because they hoped to pass it on to offspring. Yet prejudicial banking kept Black ownership rates in the first half of the twentieth century lagging behind. In 1935, the Reverend James White dreamed how the little Marshall Heights village would become one day "a thriving neighborhood, owned, controlled and managed by his people." To outsiders, Marshall Heights just looked like a "shantytown in the woods," but it would be hard to overstate what the community meant to the people who lived there.

After a long day of work, feet tired, a domestic worker would return home, climbing the hill slowly now that no one was telling her to hurry up. She would run into a neighbor who would address her respectfully by name and title. That felt right. In other parts of the city, she never knew what she would be called. She might change from a uniform into her good clothes to attend evening bible study. Home in her neighborhood meant she was no longer invisible. As Alexander Davis explained it, "If you left Deanwood, you were a stranger everywhere else."

What shines through in the stories people told of their lives is that they did best when they collaborated. In the overlooked corners east of the Anacostia River, people put into action the anti-colonial and anti-capitalist ideas of the 1930s. As they did, they cultivated what sociologist Helen Fein once called a "universe of obligation." At a time when the state and society were hostile to them, they created tiny, local commons that gave them a measure of self-determination, freedom, and food sovereignty. And that brought into view a horizon of possibilities and an uncommon prosperity.

The DC census shows a remarkable fact. Among Black residents of Deanwood and Marshall Heights, 59 percent were homeowners in 1940—almost double the average for white residents in the rest of the city (33 percent). The records also show Black people took better care of their homes. A 1934 survey found that blocks with blighted housing in DC were more likely to be white than Black.

Black residents East of the River managed to buy their homes amid the Great Depression when millions of white Americans lost theirs. How did they achieve that? It was not because they had jobs. During the Depression, Clarence Turner emphasized, "You couldn't *buy* a job." Nor did Black Washingtonians get much help from fed-

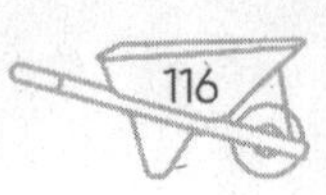

eral programs that offered white Americans affordable insurance and home loans. With no help from the state or city, residents pooled their resources and recycled nutrients to grow their food, selling some of it for cash to make mortgage and tax payments. Despite the fact that they were boxed out of the benefits of the emerging New Deal welfare state, residents East of the River obtained the American dream: homeownership in a community that welcomed them.

Purifying the Wartime Garden

PROGRAMS AND RESTRICTIONS that left Black Americans to fend for themselves in isolated neighborhoods were just the kind that Adolf Hitler admired. When the Nazis came to power in 1933, they fixated on implementing a kind of racial and vegetal rewilding of the German landscape in order to "return" native plants, native soil, and native people to Germany. To carry out his eugenic vision, Hitler insisted that Germans cleanse Germany of non-Germans, but since Germany would still need more living space, Hitler plotted out how the German Army would invade East Europe, eliminate the undesirable people there, and settle Ukraine, Poland, Belarus, and Russia with colonies of German farmers cultivating German plants and animals in the German way.

In these politics of "blood and soil," Nazi leaders applauded urban workers digging in allotment gardens and heading to the countryside in live in agricultural settlements. "As long as the German man is con-

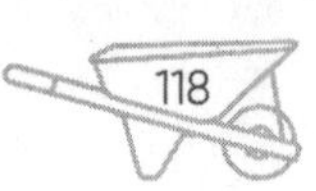

nected to the soil, Germany will live," Nazi propagandists preached. While Germans should be rooted, Jews, they taught, were rootless. Right-leaning communards called "Life Reformers" spurned artificial fertilizers as "Jewish"; the cabbages grown with them were "degenerative and sick." As they built a war machine, Nazi leaders introduced racial hygiene courses to teach citizens how to identify Jews and "degenerates." The Nazis were obsessed with the idea of purity. In botany courses soldiers learned to discern non-German plants so as to uproot and replace them with native, German flora. The Nazi leader Hermann Göring went so far as to try to back breed prehistoric tarpans (wild horses) and aurochs (cows) in the Białowieża forest of occupied Poland, believing that the closer they got to the extinct breeds, the stronger the animals would be. The projects of plant, animal, and human breeding were closely intertwined. Heinrich Himmler, tellingly, intended to turn Auschwitz into an experimental agricultural breeding station after the last extermination of Jews and undesirable Slavs.

For decades after World War II, people tried to figure out how a "civilized" European nation like Germany could carry out the Holocaust. Some scholars blamed it on historic anti-Semitism, while others delved into the psychological nature of demagogues like Hitler and Mussolini who were so good at manipulating media to bend the masses to their will. In the 1990s, the sociologist Zygmunt Bauman suggested that the Holocaust was not an aberration of civilization, but one of its possibilities. As modern states seek to conduct business as rationally and efficiently as possible, Bauman wrote, bureaucrats come to view society as a collection of so many problems to be solved, or as nature to be controlled, mastered, and improved. Modern governments thus were "Gardening States." They treated society like a garden to be shaped by lavishing care on desirable plants while eradicating unwanted weeds.

Yet inside the Nazis' Gardening State there existed gardeners' states, tiny islands of independent political and productive life. The gardeners' states were so quietly powerful that they became one of the Nazis' first targets.

In 1932, the Nazi party won the largest share of votes in parliament. Professional Germans, like the landscape architect Leberecht Migge, flocked to the Nazi Party, seeking jobs and proximity to the emergent fascist party that promised to make Germany great again. But working-class people tended to vote for the German Communist Party or the Social Democrats. Left-leaning Berliners congregated in the labyrinth of informal arbor colonies circling Berlin. "The whole neighborhood was red," Herbert Töwe remembered about the arbor colony where he grew up. The Depression drove more working families to squat illegally in arbor colonies. Nazi leaders and police, who leaned to the Right, were critical of the colonies, what they called "Negro villages." To keep the police from knowing they were living in the gardens, people darkened their windows at night and did their best to evade the state.

On the night of January 18, 1932, Werner Schulze, a major in the SA, the paramilitary wing of the Nazi Party, led 150 new recruits in an illegal torchlit march into the sleeping Felsenecke arbor colony in Reinickendorf on the northwestern edge of Berlin. City police accompanied the Nazi marchers, but at the entrance to the colony, the police truck could drive no farther on the unpaved road, and the SA marched on alone. It was not clear why Schulze bothered to go into the colony. Later, a witness said the Nazis were there to exert "moral pressure" on the communists. Another testified that Schulze announced they would hunt down communists and kill them. As the Nazis marched along the narrow garden paths, a dozen members of a communist fighting battalion hid in the grass. Some SA troopers came across one of them, Fritz Klemke, who wielded a pipe, shouted at the SA men, and fled. Several troopers ran after him, caught, pummeled, and kicked him. One pulled a gun and shot Klemke, father of three, dead. An exchange of gunfire followed. In the pitch dark, witnesses heard pounding feet, grunts, shouts, and blows. After the melee, an SA trooper, a fifty-eight-year-old art teacher, was also dead from a knife wound.

The Berlin city archive holds records of a police investigation of the murders. Detectives spoke to scores of eyewitnesses from the SA battalion and the colony, but they considered the testimony of arbor colonists to be unreliable because the gardeners "were all connected." The prosecutor charged a dozen colony residents with the murder of the SA man, and improbably, with the shooting death of their neighbor Klemke. No one from the SA was accused of wrongdoing.

When the Nazis seized power a year later in March 1933, they made the elimination of communists a state priority. Arbor colonies were a sure place to find them. In May, Nazi leaders in Berlin announced they would no longer tolerate those "little Moscows," the arbor colonies that they saw as full of "criminals and terrorists." In August, the new regime passed a law demanding that all public organizations sync up with Nazi politics and values. A Nazi party faithful took over as head of the Reich's Small Garden Association and instigated a purge of allotments. He replaced elected association leaders with Nazi members, who abolished the gardeners' councils, community arbitration courts, day care, and youth activities. One-man rule supplanted community self-governance. The swastika replaced the gardeners' green banner.

The Nazis found it easy enough to Aryanize officially endorsed Kleingärten associations, but the informal arbor colonies put up roadblocks to state control. To get a fix on the dynamic and unstable population that lived in the gardens without formal registration, Nazi officials sent out census takers, but they ran into a wall of silence, most people refusing to answer their questions. It took an inspector three months to learn the names of just three people in one colony. Surveyors estimated that about 150,000 people made the arbor colonies their home year-round. They characterized the population as poor, unemployed, uneducated, and of "low-quality stock."

Yet the colonies, like East of the River neighborhoods in DC, had most things a family needed. People ran businesses from their homes. They set up butcher shops, libraries, speakeasies, and stores selling sweets, dry goods, hardware, and secondhand goods. Colonists, a Nazi investigator found, liked living in the arbor colonies because they could cultivate a small plot of land, sell produce between jobs, and live cheaply. And, as one police observer noted, in the cozy garden communities, alienating urban life "gained meaning again." Even if their fortunes improved, many workers chose to remain in the off-grid colonies. When Nazi housing officials offered arbor residents subsidized apartments, most people turned them down. They had to be forced with "coercive measures" to leave their shantytowns.

Communists and socialists had long used arbor colonies as safe havens for political activism. After the Nazis took power, they kept up their opposition to fascism. They used the colonies to shelter people on the run from the law, to stash weapons and illegal printing presses. Gardeners made it clear that they did not accept Nazi rule. They put on puppet shows that belittled Nazis. They pulled down swastika flags. They failed to show up for meetings and refused to let their children join the mandatory Hitler Youth. They shunned gardeners who put on the brown uniform of the Nazi Party. Brawls broke out between socialists and national socialists. In addition to communists and Jews, Nazis targeted gypsies and homosexuals, but in the arbor colonies, gypsies still stopped at the water pump to sell housewares, while colonists had to be reminded with warnings to boycott Jewish-owned pubs and cafés. In an arbor colony in the Markisches Quarter, Rudi Schosstag, a gay hairstylist, organized an annual cross-dressing costume ball. In the Heinze colony, a man named Thomé printed up anti-Nazi slogans and glued them to walls and fences. Police searched the colony for the typewriter and the culprit, but no one gave up Thomé or his press.

Nazi leaders did not like them, but they needed arbor colonies to house and feed valuable workers. They permitted year-round res-

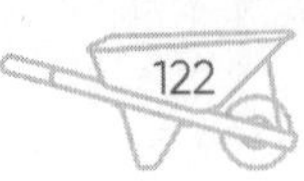

idency in Berlin's garden houses built before 1935. Working around this regulation, gardeners constructed houses on holidays when police were off duty and then insisted to police they had long been standing. When inspectors set about to demolish unauthorized dwellings, colonists threatened them. When inspectors returned with armed police to pull down the huts, residents waited until the authorities left and used the same materials to put them back up. A Nazi police chief complained, "Gardeners in no way take the decrees of the building police seriously."

One way to clean a house is to demolish it entirely. Albert Speer, Hitler's chief architect, drew up grand plans for a massive urban renewal project for Berlin. Part of the effort was to turn the "green slums" into orderly residential neighborhoods with water mains, sewers, pavement, and residents "who were true Germans." These new neighborhoods would be priced for the middle class. Working-class gardeners would have to go. Planners sought to move people with low earning potential to the countryside to subsistence settlements. In 1933, Migge proposed a ring of such settlements circling Berlin, where each working-class colony would be huddled next to a "sewage farm," vast fields of sand surrounded by drainage ditches that absorbed waste from Berlin's toilets. Migge anticipated that over a million people could leave the city center to toil on city-owned farms that used sewage to grow food for Berliners. Rather than send armies abroad, Migge insisted, "Berlin could colonize itself."

Migge died before his plan could unfold, but other unsavory settlement schemes cropped up. In 1936, urban planners created a gated garden community in the city of Bremen where they sent "anti-social" families to live and grow food on a trial basis under SS supervision. At the end of the test, those determined to be socially fit could move into subsidized public housing. The work-shy, reckless, or inefficient were sent to live in the simplest huts, where they were left to their own fate without state support. As in New Deal America, the emergent welfare state came with ideas of selecting, culling, and breeding the national population.

The Bremen project involved eighty-four families, a small venture compared to the massive, do-it-yourself experiments in sustainable settlements in Berlin's arbor colonies. Despite laws banning new "wild" or unplanned arbor colonies, the informal settlements grew in number, spreading across the city in the misery of the Great Depression. As in American cities, they were needed. The Nazi regime in Berlin had no other place to house 150,000 illegal squatters. And the produce they grew in their gardens was also useful. At the 1936 Nazi party conference in Nuremberg, Hitler announced a four-year plan to make Germany independent of imports of raw materials and food. In the "battle for production," gardeners could help. Nazi agronomists set up courses to teach people about fertilizers, high-yield seeds, animal breeding, grafting, and how to cook from a local diet. They encouraged allotment holders to can, pickle, and distill fruit into wine, cider, and schnapps. Provincial officials tallied the millions of kilograms of fruits, vegetables, eggs, meat, and honey that Berlin gardeners produced each year. From 1934 to 1937, the volume of allotment garden produce in Berlin doubled. Nazi officials were so delighted with the bounty, they planned to create 100,000 more allotment gardens.

Once the war began, with the German army marching across Europe conquering country after country, the food Berliners grew became ever more valuable. In 1940, food shortages became a problem, and the state granted citizens access to vacant lots, parks, and playgrounds to cultivate. They established quotas for allotment holders. Gardeners were forced to plant intensively to supply charities and hospitals. They were tasked with caring for mulberry trees to feed to silkworms to supply raw material for army parachutes. As the war grew more costly, the state taxed gardeners for everything they grew. Yet even as Nazi overseers exerted more control, the gardeners' state retained a certain impenetrability. Of the 1,700 Jews who survived in wartime Berlin, a good number sheltered in arbor colonies. It is hard to know just how many. Hans Rosenthal hid in the Lichtenberger arbor colony from March 1943. Max and Gerta Naujocks gave shelter to the Weiss family in the Wiesenhöhe colony. Gerda Thiess and her daughters survived living with a neighbor, and Herta Bräutigan remained in her colony, where she would walk into the small grocery store wearing a Star of David to leisurely do her shopping.

Allied bombs began to rake Berlin in 1941. Bombing raids drove middle-class Berliners to the green margins of the city in search of food and shelter. Party leaders issued guidelines for erecting huts and gardens, the same cheap, quick designs that colonists had long mastered. The dwellings were supposed to be temporary, but after the war, Berlin remained a city of rubble for years. In 1948, East Berlin registered 35,000 households (about 90,000 people) residing in garden huts. In 1954, 21,000 families (about 60,000 people) in West Berlin still lived year-round in gardens. West Berlin officials worried that colonies were prey to communist infiltration. Officials in East Berlin legalized

Children going to school before beans growing as a green window shade in a postwar "green slum." Berlin-Wittenau, 1955.

year-round residency in arbor colonies, but they fretted that the tight-knit communities secretly harbored capitalists.

On a warm June day in 2023, my research assistant Rosa and I met with a group of former East Berlin gardeners who were keenly interested in the history of their garden association, founded in 1928. The gardeners described how important the food they grew was until the fall of the Berlin Wall in 1989. East Germans relied on their plots for the fruits and vegetables that mechanized collective farms had trouble producing. When I asked them how much of their food came from the allotments, a woman in the group answered, "All of it." In her large garden, she grew potatoes, squash, beans, and fruit.

Another gardener described a cherry tree on his plot. He and his daughters would pick the cherries and sell them to the state grocery store. Laughing, the man explained socialist economics. "The state store bought the cherries for higher prices than they sold them." After the sale, he would walk around to the front of the shop and buy back his cherries, pocketing the difference.

And so it was, from monarchy to republic, republic to fascism, fascism to socialism: Through it all, small gardens were a permanent fixture of Berlin's landscape. Crisis after crisis, Berliners relied on them to survive. In 1978, an East German Berliner told a reporter that he had started his urban garden in 1918 during the final days of World War I. His garden fed his family through the hyperinflation of the 1920s, unemployment during the Great Depression, the bombs falling in World War II, and the hunger winter that followed. He self-provisioned through the meager consumptive choices of the East German socialist economy. Like thousands of others, the man had persisted, umbilically attached to his plot in the gardeners' state.

My shovel hit the earth with a ka-chunk. Wesley and I were digging trenches to lay down a line of asparagus. We didn't get far. The ground on this stretch of urban territory in Washington, DC, consisted of a thin layer of grass covering gravel mixed with a burnt-orange clay. Until a person starts digging, the depletion of the ground underfoot goes unnoticed. Yet soils spread out beneath us as important historical artifacts. Conservationists, economists, nationalists, and environmental historians have written a great deal about soil quality and the "rape of the land." For a long time, scientists assumed that soil was a collection of mineral elements. Instead, in recent decades molecular biologists have shown that billions of microscopic organisms dwell in healthy soils to form a superorganism of wondrous complexity. As I dug through rock and clay, I puzzled over how a history of this soil microbiome would read. Is it possible to populate the ground beneath my feet with the story of the tiny lives down there?

I looked under my spade. I had little to go on. There was almost nothing alive on this patch of land beside the neighborhood elementary school. The site had been a staging ground for a major construction project. For two years, bulldozers, trucks, and cranes crawled back and forth, scraping and removing plants and earth, compacting and pulverizing the ground. After the school renovation was complete, workers rolled out nylon landscaping nets, dumped a layer of commercial topsoil, sprayed some liquid nitrogen fertilizer, and broadcast grass seed. It was good enough for a lawn, but not for a vegetable garden.

Wesley and I wanted to see how many edible trees, shrubs, and cover crops we could coax into life on the heavily trafficked land around the school. The Covid-19 quarantine was going on, with its supply chain problems. Clearly, our food system faltered in emergencies. Wesley and I did not think we could solve these problems with our small project, but we wanted to know, just for the challenge of it, how much food we could grow around the edges of a city block.

We had no budget, no staff other than ourselves, and just a wee bit of permission from the school garden coordinator, who managed a small garden plot for classroom instruction.

As spring warmed the ground, we began to plant. There was no one around to take notice except for a pack of joyful children unleashed from classrooms. They paid us no mind.

Once started, we had trouble stopping. Wes and I sowed kale, spinach, arugula, and wildflowers along the fringes of the school community garden, then moved over to the east side of the building to plant strawberry and asparagus. We dug in hardy berry bushes and fig trees along the school's south facade, which bakes as hot and dry as any Italian olive grove. We climbed over the padlocked fence to slip in native pawpaw and persimmon trees in the courtyard, while thumbing into the soil seeds of corn, beans, and squash. These new edibles took root among struggling imported plants that the landscapers had laid down like carpet.

Wesley grafted fruiting apple branches onto ornamental crab apple and nestled tomato starts among decorative shrubs. We hoped the strawberry plants would spread as a cover crop to replace the English ivy where great clouds of mosquitoes bred. A city tree cutter gave us freshly cut white oak logs. We inoculated them with shitake mushroom spawn and stowed them in the shade of holly trees. Finally, we moved on to the marginal land on the west side of the school, where the cranes and trucks had done the most damage.

There, the soil had nothing to it. I dug in and lifted my spade to reveal a terrain of mass extinction. The ground held no worms, bugs, roots, or fungi. Nothing. We unearthed only "foreign" elements—glass, plastic, asphalt, brick, rusty nails. For a historian, it comes as no surprise that we live on ruins left by our predecessors, but I was in no mood for an archeology expedition. My eye was on the future, building soil and restoring dead ground to life.

Compacted ground aerates poorly and has trouble both holding water and draining it. We could see that even grass and invasive weeds had a hard time growing on the west side plot. Healthy soils are springy, lined with hair-thin roots and densely populated with a world of life in the form of insects, tiny animals, bacteria, viruses, and fungi. Specialized bacteria ally with root tips, which serve as plants' command center. *Azotobacter* produces growth-stimulating chemicals. *Rhizobium* and *Glomus* work with roots to reduce disease in soils. Another bacteria, *Streptomyces* (the same helpful bacteria in human antibiotics), rescues thirsty plants during droughts. If drought conditions continue, root bacteria can even mutate to alter their metabolism to help plants along. Other bacteria and fungi quarantine plants from insects and pests. Not all bacteria and fungi are beneficial. Some cause disease. Others are no more than freeloading parasites. It's like any community—we get the good and the bad.

Wesley and I debated what to do. We had neither the money nor the desire to buy sterile plastic bags of topsoil excavated from somewhere else. So far, we had nourished our seedlings with leaf mulch and compost from kitchen scraps. That meager stream had run dry.

Our problem of ravaged soils is one of the oldest in the history of agriculture. I doubted we had enough nutrients for our new trees and crops to thrive. I read a study about a successful and affordable soil supplement, Bloom, that was made at the city's wastewater treatment plant. Using high temperatures and pressure in an oxygen-free environment, the plant cooks Washington sewer sludge and produces a fertilizer that April Thompson, the plant's marketing director, told me is clean enough to be labeled organic (though it does not bear that label for reasons I will explain below). Unlike most American cities of its size, DC has little industrial waste in its sewage, and that makes Bloom cleaner than the usual "biosolid," which is the waste industry's manufactured term for sewage that is cooked at high temperatures to kill germs for application on farm fields.

I suggested to Wesley that we use Bloom to enrich our crops. Wesley wasn't so sure he wanted to farm with what Washingtonians flushed.

A lot of things end up in our drains—cleaning fluids, condoms, pharmaceuticals, chemical toxins, oil runoff. Human waste, Wesley pointed out, has come a long way since the more biologically innocent night soil days when gardeners East of the River used composted waste directly from outhouses. Wesley preferred to experiment with the biodynamic method—that is, the soil enrichment program Rudolf Steiner's followers developed in Germany based on a dozen strange lectures on "vital matter" that Steiner gave in 1924. We decided to try both methods.

A dump truck left us a huge, smoldering pile of Bloom cut with wood mulch. It gave off the pungent smell of ammonia and chlorine. I plunged my shovel in. Steam floated up, wafts of humid, sauna-thick air. Millions of bacteria in the pile were working away, getting heated up to as much as 140°F as they dined and burped carbon. Even through my glove, the pile was hot to touch. For days, we carted the steaming compost to tree beds. Bloom covered eroded ground of gravel and clay, but as fine particles coated my throat, I began to feel I would like a good bit more social distance from my community's waste.

Wes ordered a horn stuffed with manure that had been buried for the winter on a biodynamic farm in Eastern Maryland. Following instructions from an online course, we waited until the last quarter moon, dumped the horn's few ounces of manure in a bucket of water and stirred vigorously by hand for half an hour. We placed the brew in plastic aerators and misted it over our garden beds. Compared to hauling Bloom, biodynamic farming felt like a sunny day at a spring fair. But how, I wondered, could that tiny amount of misted cow manure fertilize our crops?

The biodynamic cocktail we made took microbes from a cow's gut and brewed it in dirt from a local farm. As we stirred, microbes from our left and right hands (they are different) joined the mix. These particular microbes formed the mysterious, "vital matter" our gardens so

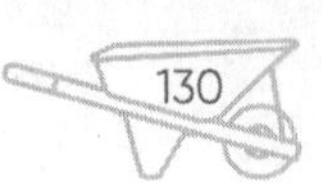

needed. The hope was that the brew contained *autotrophs*, bacteria that function like the city's wastewater treatment plant in breaking down organic materials to produce soil rich with nutrients.

We were careful to use the biodynamic method on row crops, which we and our neighbors would eat directly. We saved Bloom for perennial trees and berry bushes, which, if they took up heavy metals or chemical toxins from Bloom, would distribute them across the woody plants. Then we waited to see what happened.

I was secretly betting on Bloom. Biosolids could close the metabolic rift that broke open in the nineteenth century when urbanites began to toss organic nutrients into waste streams instead of returning them to farm fields to become nutrients again. Farmers using biosolids could save landfill space and stem the overflow of fecal matter into rivers and ocean deltas where they cause fish-killing algal blooms. Successful biosolids could replace chemical fertilizers that require a lot of fossil fuels to make. But as we spread Bloom, stories broke in the newspapers that typical biosolids harbor forever chemicals, such as polyfluoroalkyl substances (PFAS), that poison soil, produce, animals, and humans. There had to be a better solution. Humans concentrate organic waste in cities. If waste is used for urban farming, then both fertilizers and food could aggregate in urban areas where the majority of humans on earth live. For centuries, farms and gardens on the urban perimeter of Chinese cities had bountiful fertility thanks to night soil. Beijing only adopted flush toilets in the 1990s. The lesson for me was that food doesn't have to travel, and neither do fertilizers.

After planting, we stepped back and watched. The plants treated with Bloom grew swiftly at first, but, as the heat of high summer rolled in, the soils dried up, and a hard crust formed on the surface. We had to chop the ground to pry it open to water. In the biodynamic beds, the soil was spongy and held water better. In both kinds of beds, our plants grew.

One evening after hauling Bloom to tree beds, I headed home, taking a shortcut behind the school. Squeezing around a fence, I hopped over a broken PVC pipe sieving water and scrambled up a steep slope to a weedy bioswale. The sun was setting behind me. I tasted a sweaty brine washing down from my brow. I was satisfied with my evening's work. I stopped to listen to a solitary warbler pipe her call. Gold flashed across the bark of monument-sized white oaks in Rock Creek Park. The sky settled in around them, cerulean and infinite. Blackberry seedlings grew to my right, and on my left, young sprouts of kale, amaranth, and chard. I breathed in the moist air. Here stood our little farm that few would recognize as a farm. It was more like Darwin's description of an "entangled bank clothed with many plants of many kinds, with birds singing on the bushes." For a moment, I was filled with a sense of belonging, overtaken with the saturating joy of grasping a place—this city block—with a shrub-by-shrub intimacy that I had not known since I was a child.

Looking around at our work, I was struck by how the job of regenerating the soil wasn't our work at all. The plants grew, and then dropped their leaves. As organic matter piled up, worms, fungi, and microbes came and broke it down. The beds turned a springy, dark brown. It was a kind of resurrection that doesn't require faith in miracles.

11 The War at Home

DURING WORLD WAR II, people East of the River in DC had to fight to keep what space they could, as each month during the war 10,000 people arrived in DC to take up jobs in the riotously expanding federal government. With their dusty suitcases, newcomers brought a major housing crisis, especially among Black Washingtonians. Private contractors constructed tens of thousands of apartments for white war workers but only a few hundred for Black workers. Black migrants crowded into either alley dwellings or housing in areas free of racial restrictions East of the River. There homeowners continued to grow food in their gardens, and twenty million Americans joined them in self-provisioning. The US Department of Agriculture promoted Victory Gardens to ease wartime food shortages. Initially, USDA officials thought that Victory Gardens in cities would be a waste of time and money. They believed urban ground was too threadbare and

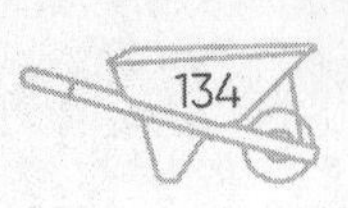

urbanites too ignorant to contribute much to the cause. But in DC, when officials lifted a ban on front yard gardens and handed over 650 acres for allotments, 35,000 people applied for plots. The Victory Gardens were phenomenally fertile. By 1944 they produced 40 percent of all fresh fruits and vegetables across the United States.

You would think that Victory Gardens would validate self-provisioning communities like Deanwood and Marshall Heights, but they didn't. Ironically, growing one's food and building one's shelter made subsistence neighborhoods suspect as economic dead space, as failure. Policymakers saw the gardens and outhouses and determined that Black Americans were unable to make the transition from country to city. Black residents, they diagnosed, had created an "unfinished city" and needed help making the full transition to urban living. Even before the war ended, DC housing reformers went looking for what they called "virgin land" in the city to build mass housing quickly and cheaply. DC leaders had long neglected Black neighborhoods East of the River, but suddenly powerful people became interested in them. Developers zeroed in on all that lush, green space in Marshall Heights.

"The present population is scattered," Alabama Senator John Sparkman observed ominously, "occupying more ground than it needs." A city planner called Marshall Heights "blighted" and pointed out that 60 percent of the neighborhood was vacant. The land, of course, was not vacant. Residents owned those lots and populated them with corn, beets, beans, fruit trees, chickens, and pigs. But real estate prospectors saw the lots just as enclosure-minded English landowners had viewed commoners' edible landscapes—as "unimproved" wastes. In 1943, housing reformers such as U. S. Grant III, the grandson of President U. S. Grant, appealed to Congress for $20 million to clear and rebuild Marshall Heights and another Black community East of the River. In 1946, Congress created the Redevelopment Land Agency as a vehicle for private developers to carry out the job of "slum clearance." In 1947, commissioners issued a freeze on new construction in Marshall Heights in expectation of wholesale demolition.

The Housing Act of 1949 established the federal government's responsibility to clear slums, assemble prime real estate, and hand it over to private developers to build with public financing. Marshall Heights was to be the "No. 1 project" for "slum clearance." Reformers had long used the District of Columbia as a laboratory for social experimentation. They found it easier to try out new schemes in a city run by decree with no fickle voters and quarrelsome city council. In DC, planners hoped Marshall Heights would show how large-scale slum clearance could serve as an effective new tool to eradicate both "blight" and "the Negro problem" in American cities. An editorialist described the revolutionary potential of replacing "slums" with attractive shopping, housing, and parks. "By implication," he wrote, "the jails will be empty, the hospitals unused." Simultaneous with the deadly Tuskegee experiments where doctors inoculated Black subjects with syphilis, reformers proclaimed that Marshall Heights offered an "excellent opportunity for a laboratory experiment," "a rare chance to study the effect of redevelopment on human values."

In Marshall Heights, residents still remember this betrayal by city leaders. Joyce Cox and her brother Bishop Jeremiah Murphy sat with me one spring day after services in the Leading Commandment Church, constructed by the siblings' father and parishioners in 1941. Sister Joyce put it tactfully. "President U. S. Grant was very kind, but his children weren't. They had their ideas." U. S. Grant III, who ran the city planning commission, determined that Marshall Heights made a good proving ground because it was "a slum as evil as any to be met anywhere." Mark Lansburgh, a wealthy businessman and chairman of the Redevelopment Land Agency, agreed. He said something needed to be done fast or all white residents would flee the city, which he joked awkwardly had become known as the "District of Co-Slum-Bia." The

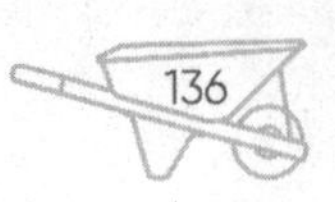

stakes were high. DC history showed that once city leaders defined Black spaces as substandard, they set out to destroy them.

What makes a place a slum? Marshall Heights residents fought hard to define themselves and their place. They said they did not live in a slum but in a family-friendly neighborhood that was developing slowly. Housing reformers admitted that they had no formal definition for substandard housing. The DC planning office assessed houses on two streets in Marshall Heights. Using the American Public Health Association's categories for crowding, maintenance, and "basic deficiencies" such as an approved water source, the two streets failed the test. Extrapolating from the tiny sample, the surveyors concluded that 86 percent of Marshall Heights homes were blighted and should be "swept away."

Crowding, poor maintenance, and lack of running water were not something that Marshall Heights residents could fix on their own: For decades code inspectors had skipped over neighborhoods like Marshall Heights and neglected to install basic infrastructure such as sewers, water mains, paved streets and sidewalks. Meanwhile, 105 racially restrictive covenants across the district continued to funnel new arrivals into the underserved neighborhoods. Black communities East of the River had no secondary school, fire station, or hospital. Without infrastructure, insurance companies would not insure housing, and lenders would not issue loans, so people repaired and restored as best they could with their limited means. If Marshall Heights was a slum, it was due to the blueprint of structural racism that had drafted it.

Declaring a "health menace," district commissioners sought to seize most of the 456 houses and eight churches in Marshall Heights through eminent domain. They admitted that some of the doomed houses were just fine, as good as anywhere in the city, but they would have to go because it was cheaper to develop the whole tract together.

The commissioners proposed to pay Black homeowners three cents per square foot, nineteenth-century prices, for their houses, orchards, and gardens. They would then sell the properties at a discount to private developers to level Marshall Heights and regrade it for public and private housing that would accommodate three times more residents than before.

Where would the new residents come from? Tearing down Marshall Heights was part of a "comprehensive plan" that was kept secret from residents but became public during Senate hearings in December 1949. In the larger plan, city officials would raze densely populated Black areas in central DC to free up space for white residents to achieve what they had long sought, a white central city. In a Senate hearing over the issue, John Ihlder, executive director of the National Capital Housing Authority (formerly the Alley Dwelling Authority) wanted to make sure they did not repeat demolitions of the 1930s, when crews pulled down alley houses with no place for the evicted to go. He proposed that new mass housing projects in Marshall Heights would serve as "overflow from congested areas." Mrs. Donald C. Beatty, president of the Columbia League of Women Voters, promoted Marshall Heights as a good "outlet" to rehouse persons "displaced permanently by the process of land clearance." Helen Duey Hoffman, of the American Association of University Women, said that in 1935 she met on a tour of Marshall Heights a single mother of nine children who had built her house herself. Hoffman admired the unnamed woman, but was sure she would sacrifice her home "so that forty others can be housed."

The plan, Marshall Heights residents pointed out, would destroy Black-owned business and "crush the Negro as a homeowner and a competitor of white merchants." The housing projects, they said, "would create a Hitler-like ghetto." A planning spokesman confirmed this suspicion. He enthused that a valley running through Marshall Heights "for which a highway is now being planned" would be "the logical dividing line between the white and Negro sections of the far southwest."

White leaders in DC sought a "reservoir," "outlet," or "overflow"—something akin to a reservation to house Black residents unwanted in central DC yet still needed as a cheap labor force. Servants, busboys, sanitation workers, janitors, maids, and nannies were jobs designated for Black people in the district. With expanded transportation networks, service personnel no longer had to live close to their employers. Rolling in on a bus, Black workers could materialize out of the far corners of the city and disappear at the end of the work shift.

Infrastructure can be deployed to organize social difference and distribute power. Infrastructure can be a form of violence. In 1957, the historian Karl Wittfogel published a book, *Oriental Despotism*, about the origins of complex societies. He described "hydraulic societies" where Chinese emperors gained power over people by setting up intricate irrigation systems that demanded coerced labor. "Those who control the (hydraulic) network," Wittfogel wrote, "are uniquely prepared to wield supreme power." A different kind of hydraulic society was at work in DC in the 1940s, one no less despotic.

Specifically, toilets became the chief lever of coercion, the justification for the formation of something that would work like the South African Bantu homelands also taking shape in these years. City planners said slum clearance in Marshall Heights was necessary because people lacked indoor toilets. Outdoor privies, public health officials insisted, were a "water pollution threat" that could result in raw sewage siphoned into city drinking water. I found, however, no recorded cases of people falling ill in Marshall Heights from contaminated water.

Privies have a poor reputation that is often undeserved. The first large, dense cities in the world were in Tokugawa Japan and Qing China, powered in large part by cycling human and plant waste through urban farms. (The water in these cities was also cleaner than in European cities at the same time.) Contrary to popular assumptions, streets of medieval cities were not awash in foul and fetid matter. In the Middle Ages, Europeans successfully managed crowded urban life without flush toilets. And so did people in Marshall Heights until the 1960s. The bucket men came weekly, emptied outhouses and recycled human waste into fertilizers in a well-functioning circulation of nutrients. People also dug composted human waste directly into their gardens.

At the same time, the city's sewer system continued to aggregate street runoff with sewage and dump it untreated into the Anacostia and Potomac Rivers. In 1943, fecal contaminants seeped into drinking water and people fell ill. Despite pump upgrades and later expansions of the wastewater treatment plant, city engineers did not manage to prevent sewage water from flowing into the river after heavy rains. DC was not an industrial city. Its major business was the federal government. The chief source of water pollution came from federal employees and residents flushing toilets. Even so, by the 1970s, the Anacostia Watershed was an ecological dead zone, the river ranking in the top ten most polluted rivers in the country.

Americans flush on average 9,000 gallons of fresh water per person a year, about 24 percent of the nation's water use. In the future, flush toilets will be a luxury only the rich will be able to afford. That is why in 2015, the prize-wining architects for a utopian Deanwood designed housing with composting toilets, the kind that residents had long used East of the River until they were condemned by the hydraulic society.

Not that people in Marshall Heights liked going down the block to get water or out to the privy in the cold. They wanted indoor plumbing like everyone else. By law, the district commissioners were obligated to provide water, sewer, and roads in Marshall Heights as elsewhere in the city. Residents had already paid assessments for sewer and water lines, curbs and sidewalks. If the city ever made good on those promises, then homeowners would need only to afford to connect the mains to their houses. Residents said they would happily assume those costs. They asked for a municipal bond to help them underwrite the expense, but the bond market was as racialized as other financial tools at the time. Besides, city planners conjectured, plumbing would cost thousands of dollars, too much for the poor people of Marshall Heights. A DC official corrected the planners. He said the hookup expense would require only about $9 a house because “a lot of people do their own work, and that reduces the cost.”

The developers and planners made use of other facts that turned out not to be facts. In order to justify wholesale demolition, they drastically understated the value of homeowners’ property and significantly overestimated the cost of infrastructure to bring the neighborhood up to standard. The Black civil engineer and activist Howard Woodson pointed out that housing in Marshall Heights was not blighted. He had toured the area with a city building inspector who calculated that not 86 but only 25 percent of the houses were in a state to be condemned. The rest of the houses were fine. In 1949, as the debate over Marshall Heights played out, a white photographer, John Wymer, took dozens of photos of the neighborhood, which he described as a “suburban slum.” Wymer snapped pictures of one shack made of found materials,

Boy drawing water from street hydrant, the only running water in Marshall Heights, March 27, 1949.

but most photos show handsome two-story dwellings surrounded by trees and greenery. His lens focused on several well-built brick churches and smaller, wooden "family" churches. In Wymer's photos, the infrastructure is what is blighted: rough, rutted streets with primitive water pumps on the corners. Standing by the pumps are children, dressed in clean clothing in good shoes with trusting grins. The children do not look like slum-dwellers.

Residents of Marshall Heights were finally allowed to testify at 10 p.m. on the last day of the Senate hearings on slum clearance, after some senators had already left. Chairman Sparkman told the Black citizens to make it quick because it was late, and he wanted to get home.

Patsy Allen, leader of the Pioneers Committee of Marshall Heights, stood up: "I think you ought to read the records very carefully before you write any more editorials telling us how we should smile freely while somebody tears down houses that we worked so hard to build with no help from you or the government or anybody else. . . . We consider ourselves pioneers because we bought and cleared the land and were the first to dwell on it. We feel that our homes are as good as anyone else. We have both comfort and recreation there. . . . Our homes are debt-free. They are ours. At least we thought they were ours until this redevelopment nightmare came along."

Allen gave the senators a lesson in the political economy of the emerging welfare state, which, in giving precedence to white Americans, tipped the scales against Black Americans.

"Our situation has been made even worse because our government has been helping those who are trying to take our homes away from us in order that certain government officials can have jobs, can spend money which they did not earn by taking from us the homes which we did earn, and in order that certain private builders can be given our land at a low price for their profit."

"If government can do it to us in Marshall Heights," Allen added, "they can do the same thing to any property owner anywhere."

"You will not want to live," Howard Woodson testified, "alongside a public housing project yourself." White and Black neighborhoods had already campaigned against housing for low-income residents in their neighborhoods. Woodson predicted that he and other middle-class Black residents would move out of Marshall Heights when public housing arrived.

Ernest Harper, from the National Association of Colored People, pointed out that the clearance plan would transform property owners into precarious renters. "If people are moved out, how can they secure a down payment and loan for a new house? The effect of past plans has been to take away these homes, thus leaving such persons no place to go." "Thousands of Negroes," Allen observed, "will lose their life savings and become renters or objects of charity." The elderly, in particular, had no place to relocate but to Blue Plains, the city's miserable, underfunded home for the destitute aged.

In the hour allowed to them in the darkness of that December evening, the people of Marshall Heights made their case powerfully. They would not back down, they promised; they would fight for their homes as did the "Pilgrims who landed on Plymouth Rock in 1620." They called on the American spirit of self-reliance and they spoke of the sanctity of private property.

"When my husband and I bought our land, it was in the woods," Patsy Allen observed. "Now we have our home, and these hands of mine helped build it. Is that an unforgivable sin in America?"

For seven years, the people of Marshall Heights made their case, signed petitions, appealed to Congress, and generally defied the plot to turn their part of the city into a "Black ghetto." Moved by their testimony, Congress refused to fund the demolition. In 1950, DC housing officials admitted defeat and agreed to call off Marshall Heights "slum work."

Unfortunately, there is more than one way to strip people of their property. Soon after the commissioners halted plans for the whole-

House in Marshall Heights designated for "slum clearance." 1948.

sale clearance of Marshall Heights, the Federal Housing Administration (FHA) set up an advisory committee to "fight blight." The FHA appointed six national leaders, wealthy giants in real estate, construction, financing, and insurance. The Baltimore developer James Rouse, "a national figure in slum clearance," led the committee.

Developers understood "blight" literally. In a mono-crop field, a fungus infects crops and spreads quickly, killing plants. Developers believed that neighborhoods worked the same way. Each residential area was a vulnerable, mono-class, mono-race field. One patch of blight—say a lot with tall weedy-looking plants or the arrival of a Black family—caused a contagion that would spread from one house to another, destroying the neighborhood. Developers considered blight inevitable, a function of aging and race. The trick was to stay ahead of it. They imagined a continual renewal of housing to stave off "slums of the future." This was a profitable imaginary for those in the construction and real estate industry.

Since Congress ruled that they could not demolish Marshall Heights wholesale, Rouse determined they would take it down piecemeal by using "law enforcement machinery equipped with a housing code." The mechanism worked two ways.

First, city crews arrived to install long-awaited infrastructure in Marshall Heights. This should have been cause for celebration, but new infrastructure turned out to be punitive. After paving, for example—not a street, but an alley—city officials handed bills to residents for $600, a quarter of their average annual income. Those who didn't pay were fined. When district engineers extended sewer and water mains into Marshall Heights, inspectors gave homeowners thirty days to hook up or face court action. Families scrambled to get the money. A third of homeowners could not afford to connect, and they were ticketed. One man could not pay a $15 fine and spent several nights in jail. A public health official noted that for residents, "The financial strain is terrific." Those who could not pay to hook up or afford fines had their houses condemned.

The second method Rouse used to clear Marshall Heights involved sending out building inspectors to make frequent house-to-house searches for code violations. Inspectors descended on Marshall Heights each year. They would enter houses, look around, and find breaches in code in over 80 percent of the properties. The head of the civic association explained that it was difficult for Black residents to obtain home improvement loans for the same redlining reasons they could not qualify for loans to purchase property. The fact that Marshall Heights continued to appear on maps as slated for removal made getting a bank loan doubly difficult. By 1953, city inspectors had condemned seventy-five houses in Marshall Heights as "a grave health menace." As houses stood vacant, with windows broken, Marshall Heights became a place where people dumped old cars and appliances. Blight, indeed, spread like a contagion.

Across the country, wrecking balls were hard at work, pulling down 7.5 million dwelling units nationwide between 1950 and 1980. Bulldozers crossed the country to take down neighborhoods in every state from Miami to Seattle to Little Rock, from Detroit to Dallas and Denver. In cities large and small, bulldozers engaged in a battle against blight. They pushed over houses, scraped up the earth, leveled it, dug trenches, poured concrete, and paved. Miles of asphalt and pavement replaced fertile gardens, orchards, and groves of nut-bearing trees.

Punitive infrastructure, draconian building-code enforcement, and demolitions worked to clear space to build the long-desired reservoir of Black housing East of the River, while city planners moved on to their primary goal of "revitalizing" the southwest quadrant of the city between the US Capitol and the Potomac River. In 1957, Congress approved an enormous sum, $500 million (equivalent to $5 billion in 2024), to bulldoze nearly the entire southwest quadrant. The project displaced 23,000 residents, most of them Black. Crews built new, modernist high rises, wide arteries, stone plazas, shopping centers with large parking lots, and reflecting pools for 10,000 mostly white residents. Developers had promised to construct low-income housing in the area, but they ran out of energy for all but one apartment building. Nearly half of the evicted moved East of the River, where construction crews hastily put up six large public housing projects, doubling the number of dwellings in Deanwood and Marshall Heights.

As thousands of new people flooded in, the neighborhoods changed.

With urban renewal, neighborhoods East of the River became a kind of junkyard for unwanted citizens. Middle-class Black residents moved out of areas packed with housing projects. To qualify for vouchers in public housing, mothers could not have their children's father live with the family. Inspectors came around to ensure only those on the lease stayed the night. Social workers asked snooping questions, suspicious that welfare recipients were cheating the system. Public housing regulations not only pried fathers apart from their families but ended a long-standing practice in the community of hosting family and acquaintances in need of shelter. The number of people living on the streets, once a rarity, grew in DC decade by decade. Service jobs, those long reserved for Black residents, dried up. Unemployment among Black youth in DC reached 50 percent.

City engineers, already accustomed to using the area as a sacrifice zone, moved the city dump across the Anacostia River, where sanitation workers, many of whom lived in Marshall Heights, burned solid waste. New postwar materials such as plastics and vinyls blazed into a thick, black smoke that coated a person's throat. People East of the River were used to having their environments move through them in the form of the food they grew and the fish they pulled from the river. The presence of acrid plumes of burning trash was visceral.

"Poison is the wind that blows from the north and south and east," the musician Marvin Gaye sang. "Have mercy father, fish full of mercury."

Gaye grew up in a housing project East of the River known colloquially as "Simple City." Gaye, who released "Mercy, Mercy, Me" in 1971, knew well the dirty river water and the pungent taste of smog coming from burning garbage in the nearby dump. After artificial fertilizers became cheap, sanitation workers stopped composting human waste, and they deposited the once-valued material into giant piles of feces along the Anacostia River. People fishing in the river waded through trash and pulled out bass with sores and ulcers. In the war on the environment, poor urbanites stood in the first line of fire, their bodies taking the blow. In the postwar decades, neighborhoods East of the River scored at the bottom of the city for air quality, access to green space, and fresh food, while topping the charts for chronic diseases and short life expectancies.

It is a strange experience walking in Marshall Heights. The sidewalks run along for half a block, then trail off into a grass path. Jorge Lopez explained that he filled out online forms for many months to get a sidewalk in front of his house. Apparently, only squeaky wheels have sidewalks. In 1975, a diverse group of residents unsuccessfully sued the

city for "discrimination in the provision of municipal services including policing, firefighting, trash removal and provision of sidewalks." In the 1980s, city crews installed walkways in half of Marshall Heights and then mailed homeowners bills for $900, even though a federal block grant had already fully paid for the pavement. That adds city services to the tally of costs for which Black urbanites were overcharged.

Public housing projects presented as a solution for "substandard housing" only led to more substandard housing and more demolitions. Cinderblock apartment complexes were cold, damp, and cracked. In 1988 the Deanwood Gardens housing project was already so seriously run-down that the city considered replacing tenants with prisoners. Eastgate Gardens in Marshall Heights was boarded up in 1992.

Curiously, while newer apartment blocks collapsed, older houses that Marshall Heights residents built remained standing. The fact that these houses were not condemned is a testament to residents' stubborn refusal to be displaced. In the 1980s, the Marshall Heights Community Development Corporation acquired worn-down housing to renovate for low- and moderate-income families. Property, no matter how modest, mattered deeply to the community. DC councilwoman Willie Hardy urged her neighbors to "Keep that house or shanty or anything else in Marshall Heights. You'll never find a place like this." In her novel *Song of Solomon*, Toni Morrison expressed a similar sentiment: "Own it, build it, multiply it, and pass it on—can you hear me? Pass it on!"

Pass it on they did. Records in the DC Recorder of Deeds show that for nearly a century many houses in Deanwood and Marshall Heights changed hands for $10. People deeded their homes to family members and friends for the minimum sales price rather than sell on real estate markets that overcharged Black people for financing, brokerage fees, and taxes. Norman Dale described how his family acquired their home from an elderly woman with no children.

"She didn't want to sell it to white people. She wanted some Negro to have it. She asked my father if we would take her and keep her for the rest of her life. We did."

Urban renewal functioned as a form of enclosure. Much as in nineteenth-century England and twentieth-century Berlin, it stripped working people of property, gardens, and means to self-provision while adding expensive infrastructure and taxes that pushed people into exploitative markets they could ill afford. Many people ended up in public housing and on food stamps in communities that spiraled into dysfunction.

But from 1860 to 1950, Black residents East of the River had used access to land and cooperative arrangements to selectively free themselves of capitalist markets that punished them. This was a brilliant strategy. Market-shy behavior such as bartering, trading, sharing, subsistence, and do-it-yourself construction saved Black Washingtonians millions of dollars. For those few who survived urban renewal, there would be a trade-off.

To keep your house, you had to gain the approval of code inspectors and insurance adjusters who calculated value using new suburban housing as the standard. Suburban homeowners did not keep chickens or have corn and tomatoes in the yard. To play by the rules in postwar America, homeowners increasingly had to get rid of their "chicken yards" and transform from producers into consumers. To do that, they needed a lawn.

THE FOURTH TURN

Lawn and Order

12

Green Privilege

IN 1998, GEORGETTE NORMAN LIVED ON Rosa Parks Avenue in Montgomery, Alabama, in a minimalist ranch house passed on from her parents. In her front yard, waist-high rye and mustard rasped the sidewalk. Sunflowers and tongues of sticky tobacco lapped at the fence. Muscadine and ivy wound up an arbor shading the front door. Norman and her companion, Nomad Steward, scavenged cement shards from demolition jobs to build a graceful stone wall decked with irises and roses. Norman laid down seeds of peas, potatoes, buckwheat, and squash. Many of the plants were hundred-year-old heirloom varieties. From photos, her garden was elegant and lush, gorgeous.

Norman, an artist, civil rights activist, and the founding director of the Troy University Rosa Parks Museum, saw the wild growth of plants in her yard as a means of self-expression. Plants speak to one another. People speak to plants. Humans use plants to send messages to society. Norman declared her yard "an act of conservation and restoration." "My yard is agriculture in the sense that I have chosen to grow edibles instead of grass. I have decided in a very small way to control some of what I put in my body."

Randy Guy, Montgomery's chief inspector for housing codes, read Norman's yard differently. Blight. Invasive, encroaching weeds served, he argued, as havens for mosquitoes and rats. "If you live in a neighborhood, you don't want a jungle next door to you." Guy did not explain why a jungle would be a bad neighbor. Instead, he cited Norman for a code violation, a long-standing ordinance determining that weedy plants "of no value" over twelve inches high had to be cut down.

The determination of what is a weed is in the eye of the beholder. A weed is any plant out of place. And vegetal literacy varies wildly. Norman waited for the code inspector to come and see the garden himself. Surely once he laid eyes on the flowers blooming, the squash creeping, the bees humming, he would see the mistake. But as Norman's sprinklers whispered, disappointment saturated the encounter.

"The zoning inspector couldn't even identify a collard green!"

"I told her she could take her gardening operation to the back-

Georgette Norman's garden, fined as an eyesore by Montgomery officials.

yard, and it would be acceptable," Johnny Redding, the city inspector, reasoned. Norman refused to chop down her plants and restore her front yard to turf grass. City officials responded by charging her with a crime.

"They convicted me in criminal court with all the whores, pimps, and drug addicts!"

Norman sought out her city councilman for help. He recommended she plant grass and mow like everyone else.

"Why cause trouble?"

"If Rosa Parks had stepped off that bus," Norman shot back, "you wouldn't have your job."

Feeling isolated, Norman hired a lawyer and appealed the $30 fine and criminal conviction. In boycotting turf grass, Norman ventured into her own Rosa Parks act of resistance.

Rosa Parks Avenue, where Norman lives, only won that name in 1988. A lot of people in Montgomery still call it Cleveland Avenue, named after President Glover Cleveland, who passed through on a whistle-stop campaign tour in 1887. In a frenzy of love for a president who supported Dixie Democrats' segregationist policies, white Montgomery leaders renamed Plank Road after Cleveland.

Cleveland/Rosa Parks Avenue ran through West Montgomery, a historically Black neighborhood with a commercial district centered along Jeff Davis Street. West Montgomery had a park Black people could safely enter. The neighborhood was studded with churches, single-story shotgun houses, and mid-century ranches. Gardens surrounded the houses on the side and out back.

"My mother *loved* flowers," Norman recalled. She had blossoms all over the front yard, while her father planted the side and back yard. "Collards, beans, squash, the usual."

Norman's grandfather had a farm in nearby Hope Hull, Alabama. "He would bring stuff up and supply food for the cafeterias." Norman's grandfather also carted produce in big baskets to his son, who was in Montgomery to attend high school. His job was to hawk the goods. "My father," Norman explained, "was the kind who was *not* going to be selling anything on the street. So, he had clients." He delivered the produce weekly to his customers. While still in high school, Norman's father invented a network of dedicated consumers of local farm goods, what today the Slow Food movement would call Community Supported Agriculture (CSA).

Her father was not alone.

"Most families had families that lived rur—aaal. You'd sit on the side porch and shuck corn, butter beans. There was enough for the whole community, pretty much. Each year, grandfather would kill a hog. We never bought meat."

By the 1940s, the whole neighborhood was redlined. Many residents rented their houses from absentee white owners, who could be neglectful. As late as 1960, a quarter of houses in West Montgomery lacked indoor plumbing. At the same time, produce streamed in from backyard gardens and nearby farms. As in DC, homegrown fruits and vegetables brought a certain wealth to people who existed in a gray-market economy.

"It wasn't like we didn't have," Norman reflected. "What we didn't have was citizenship."

Montgomery's city attorney declared she had no idea what Norman's assembly of plants was meant to express. "Although Norman may have intended to use her yard to send a public message about reclaiming her neighborhood, protecting the environment, and the state of today's society, there is little likelihood that her dispatch would be understood by those who saw three-to-four-foot seed rye growing in her front yard and close to the street."

After Norman won her case in the Montgomery Circuit Court, the city sought to overturn it on appeal. In the Alabama Court of Criminal Appeals, Norman argued that the Montgomery ordinance regulating homeowners' yards was unconstitutionally vague, overboard, violated

Georgette Norman watering her front yard garden.

her due process rights, her rights to equal protection and free speech, and her Fifth Amendment property rights.

Norman told the court that she started her garden as an attempt to reclaim Rosa Parks Avenue. In the years when Norman lived abroad—the 1970s—two interstate highways (I-65 and I-85) crucified West Montgomery with a four-lane asphalt cross. Pressed by buzzing cars, homeowners sold out. Those who stayed saw their property lose value. Houses stood empty. Fewer people walked the streets. Crime multiplied.

"My neighborhood was plagued with car-based prostitution," Norman told the court in 1999. "For years I have spent fruitless hours on the phone to 911. I felt that I needed to assume a stronger residential presence outside the house to curtail unwanted activity."

The Criminal Court of Appeals overturned Norman's ruling and convicted her of violating Montgomery's housing code. Norman was ready to take her case to the Alabama Supreme Court, but after the election of a new mayor and administration, Montgomery city attorneys sent her a letter annulling the case.

What was this conflict about? Why did city authorities spend so many taxpayer dollars to prosecute a few errant stalks of ryegrass? Scanning court records across the country, I saw a raft of cases involving homeowners who thought they had a choice about the plants they grew on their property. They were wrong. Between Norman and her garden-centered artistic expression stood one lowly, but hallowed plant—turf grass.

By the time Norman clashed with Montgomery officials, restrictive gardening laws were decades in the making. In 1962, Montgomery passed Ordinance No. 63–92, decreeing that all front yard spaces be planted in turf grass that was "neatly and orderly maintained." Turf grass is the third largest crop in the United States after corn and soy. One-hundred-and-thirty-five million homeowners and renters plant and maintain lawns, a cooperative national mass effort. Why do a third of all Americans go to the effort and expense of maintaining a crop they cannot eat or sell?

I weed-whacked my way through a shelf of books to figure out why Americans dutifully mow and treat their lawns. Humans have a deep-seated psychological attachment to the green color of grass. In the Cold War 1950s, Americans sought conformity and easily buckled to social pressure to maintain a lawn. Turf grass is a product of major marketing and lobbying efforts. Homeowners kept lawns to sustain property values. People tend lawns out of a desire to be good neighbors, to show their care. Children and adults play scores of games designed just for turf grass. Green is the color of American democracy and American dollars rolled out in a vast unbounded sea of grass, a botanical open-arm embrace.

Yet all the authors writing about lawns agree that as inevitable as turf grass appears, lawns are not a natural feature of the North American landscape. In the nineteenth century, front and back yards were spaces for work and self-provisioning. Initially, turf grass was a form of conspicuous consumption.

In the early French Renaissance, a few elites grew grass to adorn their gardens. Two centuries later, lawns got a boost when English aristocrats kicked tenants off their land and enclosed it. The rich wanted glen and dale for hunting game. They had an easier time seeing prey across mowed meadows. They hired former tenants at low wages to scythe, roll, and weed their lawns. An acre of lawn cost three workmen a day's labor. A lawn was a statement of its owner's wealth and mastery over place and people.

In the United States, lawns at first only adorned places of power. Turf grass framed public squares and buildings like the White House and the Capitol. Frederick Law Olmsted adopted turf grass for New York's Central Park after he evicted hundreds of families from green shantytowns on the upper east and west sides of the park. Wealthy Americans began to pay high prices for lawns to adorn their mansions and townhouses. Turf grass elevated real estate prices.

The lawn became a middle-class symbol after World War II. People buying into Levittown's first suburban developments were handed grass seed and signed a covenant promising to mow their lawns once a week. In the late 1940s, public health departments passed laws banning noxious weeds. Officials worried about pollen (although grass and street bank trees produce more pollen than most "weeds"). Municipal leaders followed up with ordinances dictating height restrictions for front yard plants—six or eight inches—rules that canceled out most vegetation other than ivy and grass.

Urbanites moved out of cities into suburbs, looking for the quiet of the countryside. They were horrified to find that farmers with loud machinery and baying animals lived there too. They passed anti-agriculture laws, banning farm animals and crops. As US metropolitan areas spread outward, lawns became a symbol of suburban order over perceived urban chaos and messy rural production. The lawn served as an avatar of bedroom communities' dedication to rest and leisure.

Because the cost of maintaining a lawn placed grass-grown respectability out of reach for working-class people, homebuyers used lawns to judge a neighborhood's status. In the complex mosaic of American urban and suburban territory, geographer Paul Robbins argues that lawns served as an ideology, a set of practices that reminded people of shared beliefs and values. In the 1930s and 1940s, for example, realtors visualized race and class by drawing up maps that color coded neighborhoods for lenders to determine creditworthiness. The map indicated the "best" neighborhoods with a bright, grassy green color. Red splotches ran across "hazardous" areas where racial minorities lived. The maps were confidential. Realtors did not show them to buyers. Instead, people calibrated neighborhoods by judging the quality of lawns. Plants, which are supposed to be apolitical, did the work of drawing abstract lines of race and class onto the landscape. Grass outlined in glimmering viridescence where desirable people (usually white) belonged and where undesirables did not. The Coldwell Lawn Mower Company instructed that poorly tended lawns were a sign of "shiftless home builders and owners with careless habits." Once turf grass set the standard, other people had to buy into it if they wanted to achieve respectability.

In cities, Black middle-class residents worried about propriety and what they called "racial uplift." They pressed their working-class neighbors to turn their "chicken yards" into lawns. Block clubs organized Home Beautiful contests to encourage turf grass and roses over green beans and corn. Garden clubs educated the public on how to grow turf grass. They targeted working-class immigrant and Black neighborhoods where people tended to use yards to grow food.

It wasn't an easy sell. For people used to subsistence, spending time and money on inedible grass when you could have sunflowers and potatoes made no sense. Former cotton farmers considered turf grass to be a noxious weed. In 1948, members of the Chicago Urban League congratulated themselves: "It's something of an achievement for a voluntary organization to persuade people who fought grass up and down cotton rows" to nurture it on their front lawns. The urban theorist Jane Jacobs described a housing project in East Harlem with a conspicuous rectangular lawn that, curiously, tenants hated. When asked, a tenant explained why:

"Nobody cared what we wanted when they built this place. They threw our houses down and pushed us here and pushed our friends somewhere else. We don't have a place around here to get a cup of coffee or a newspaper even, or borrow fifty cents. Nobody cared what we needed. But the big men come and look at that grass and say, 'isn't it wonderful! Now the poor have everything!' "

Lawn beautification campaigns mattered greatly because in the quest to become respectable, a great deal hung in the balance—housing values, tax rates, federal subsidies—and so the quality of vital services such as schools, healthcare, transportation, and sanitation. Unfortunately, because of decades of persistent discrimination in lending to people of color, even the greenest grass did not spell prosperity.

In postwar suburbia, turf grass became necessary because it covered over the fissures of ruined ecologies. Preparing new suburban tracts, contractors cleared all vegetation, bulldozed forests, and leveled hills. Developers created what historian Adam Rome calls "a wasteland, a barren plain of dirt." Working quickly, construction crews rolled out sod around the new housing, planted a few trees, and left for the next development. In this scorched earth terrain, builders used wartime technologies—tanks repurposed as bulldozers, chemical weapons retrofitted as pesticides, and explosive nitrates made into artificial fertilizers. As suburbanization continued, housing lots grew larger, setbacks from the street lengthened, and the size of lawns magnified. Suburbanites moved into a host of environmental problems. Water slid like frying eggs from a griddle across turf grass on compacted soils. Runoff inundated storm drains and septic tanks that flooded basements and left lawns soggy with fecal matter. Suburban soil eroded to an extent that put the Dust Bowl to shame.

As damaging as it was, by the 1960s, the vast majority of American municipalities had passed laws that one way or another left homeowners little choice but to plant turf grass in their front yards. Americans may love their lawns, but for decades, they had no choice about it.

The fiscal and environmental cost of cultivating turf grass is substantial. The US Department of Agriculture spent millions of dollars on research to figure out how to grow English turf grass in places where it doesn't normally grow. Lawn care services annually pour eight hundred million gallons of fuel into maintaining turf grass. A small lawn of one thousand square feet requires two inches of water a week. People purchase billions of dollars in seeds, sod, machinery, and fertilizers to keep the delicate mono-crop going through all kinds of weather conditions. Landowners "treat" their lawns with more herbicides and pesticides per acre than commercial farms do. The chemicals percolate not in distant fields, but just outside the door. Children and pets roll around in them. Barbecue guests track the toxins inside on shoes and clothing. Studies reveal that chemical lawn contaminants show up inside houses on rugs, couches, pets, and in residents' urine. Lawn-treatment chemicals have been linked to childhood brain cancer, leukemia, neuroblastoma, nontissue sarcomas, non-Hodgkin's lymphoma, and cancers of the brain and testes. Fertilizer floods local water sources and kicks up bouquets of algae as green as the best lawn. To keep waters clear, municipalities pour dog- and duck-killing herbicides in ponds and lakes.

What does it say about a nation that devotes so many resources to cultivate a plant that has so little utility and does so much damage? Why did thousands of municipalities come to enshrine into law and defend with the power of the police state such a sorry, sexless little plant? The more I looked into it, the more suspect the innocent tufts of closely mowed grass became.

The Vegetal Line

SCANNING CASE LAW for controversies like Georgette Norman's in Montgomery, I found only a few homeowners went to court over weed or vegetation laws in the 1950s and 1960s. But starting in the 1970s, a time of hippies, Black Panthers, Indigenous rights movements, and pot-smoking lovefests, more Americans went to court over their yards than ever before. This trend gathered force in the 1980s and 1990s. I wondered about that. Why would there be a rise in executive action supporting restrictive vegetation laws at a time of cultural liberalization? What do garden-variety plants have to do with culture?

In the 1960s, American urban and suburban communities began to desegregate at a faster clip after federal laws and court rulings ended legal segregation of schools and housing. Despite the new laws, white Americans still resisted people of color moving into their neighborhoods, but the way they did so changed. Initially, as Black families began to purchase housing in traditionally white neighborhoods, white vigilantes attacked Black neighbors for crossing color lines. White Chicagoans let loose rock fusillades against children swimming in Lake Michigan. Mobs set upon streetcars, hauling out Black riders, and white terrorists bombed houses of Black property owners. In Boston, angry white parents surrounded buses transporting Black children to their neighborhoods. In Philadelphia, police joined white residents in rioting against Black neighbors. Mobs of residents—sometimes two thousand strong—met their new neighbors not with plates of brownies but wielding sharp lawn tools.

Newspapers rarely reported on this "hidden" violence that pulsed through US cities, but locally the vengeful mobs were hard to miss. Southern editorialists pointed out that cities in the North were no better than the Jim Crow South. Headlines as far away as Moscow condemned white terrorism as the lie behind American democracy.

In 1959, in Chicago's West Garfield Park, white residents started to see the futility of violent resistance to desegregation. Gordon Mattson told the local *Garfieldian*, "Mob action and violence is no way to solve the problem. It only plays into the hands of the speculators who want the neighborhood to panic so that they will be able to make big profits in the buying and selling of property."

There had to be another way to keep undesirable neighbors out. Neighborhood associations and block clubs began to focus on zoning laws and ordinances related to health and safety. They were not racists, they told themselves, just good citizens who sought safe neighborhoods to raise their children. White neighborhood associations began to use their access to authority and police power to pursue code enforcement selectively to cleanse undesirable residents (often people of color) from their communities.

In Montgomery, Georgette Norman did not have access to city hall. When she called the police to report prostitution in cars in front of her house, the police officer responded, "Well, that's Rosa Parks, for you!" He meant that crime was normalized in her neighborhood. Law enforcement tacitly accepted illegal activity as long as it stayed in her part of town. Her calls went nowhere.

But white residents in the same years teamed up with city officials to deploy vegetation ordinances selectively as a way to curate their communities. Yard care ordinances worked in tandem with zoning law to sort by wealth, class, and race. Weed and plant ordinances tended to be extremely vague. City ordinances often did not define the terms *weed*, *ornamental*, or *useful*. Laws that leave definitions up to subjective court rulings violate the Due Process Clause. For example, a gardener in Chicago, Kathy Tully, won a city award for naturalizing her front yard with native wildflowers and plants. A few years later, code inspectors slapped steep fines on the very same plants for violation of height restrictions. In court, Tully showed photos and identified each beneficial plant in her garden. The judge fixed on the tall milkweed along the fence. Milkweed is great. It attracts and feeds endangered Monarch butterflies. The judge did not see it that way. The name said it all, he declared, milkweed was a weed and so a nuisance. In the South Side of Chicago, an inspector issued a $600 fine to Nanette Tucker for weeds over ten inches in height. Tucker's offending plants were day lilies, flowers she dug in to bring a bit of color to her economically struggling community.

Vague laws play an important role in the history of segregation. Lawmakers in the Jim Crow South passed laws about "vagrancy" and "loitering" that could be used selectively to control the movement and activities of freed men and women. These laws made it dangerous for Black people to leave their neighborhoods or even, in Montgomery in 1907, to sit on a wall. Ambiguous vegetation laws serve similar goals. Law enforcement can deploy them as they see fit, and neighbors can weaponize the law by phoning in complaints. I asked Norman how the city came to persecute her ryegrass.

"No neighbor complained," she replied. "It was coming from the city."

Her neighborhood had plenty of overgrown, untended lots and tall weeds amid tumbledown bungalows. Norman had worked with friends to clean up vacant lots and plant vegetables and flowers. Tall plants on Rosa Parks Avenue were not exceptional.

Norman suspected the police targeted her because she was in an intimate relationship with a white man, Nomad Seward. She lived in a historically Black neighborhood, and their intertwined lives broke a long history of violently upheld prohibitions against relationships that crossed the color line. It's hard to say if Norman's suspicion is true. Few officials openly admit racist motives. But Norman's case is not an anomaly. Scanning legal records with the help of research assistants, I found an uptick in court prosecution of vegetation laws from just a handful recorded in the 1960s, to several dozen in the 1970s and 1980s, to scores of cases in the 1990s and early 2000s. In 2009, fines for yards in the city of Chicago totaled $1.6 million. In the 2010s, the city began to net ever-larger large sums fining homeowners over plant height restrictions to help make up budget shortfalls. Each subsequent year, revenues from vegetation fines increased. In 2013, weed fines had tripled to $6 million.

The growth of homeowners associations (HOAs) played a role in tightening a green noose around homeowners. In the 1970s, the United States had just ten thousand HOAs. HOAs multiplied in the 1990s and continue to increase into the hundreds of thousands. In the United States today, condos are the majority of new home starts. HOAs tend to have tight restrictions on aesthetics. HOA covenants include designated lists of shrubs and trees, ever-shorter height restrictions on plants, and long lists of banned plants. HOAs carefully police these rules. As one man in California put it in 1994, "Careless, sloppy people who do not care about appearances will think twice before buying a home here." The increasing frequency of court cases over front yards and backyards shows that some people worry the mono-cropped order of American urban and suburban spaces is under threat.

In these cases, we found, people charged with growing vegetables in residential communities were more likely to be people of color (usually either Black or a recent immigrant), but not always. In Orlando, Florida, Jason and Jennifer Helvenston, a white couple, plowed up their front yard to grow vegetables to save money. "A budget thing." The city fined them $500 a day until the garden was replaced with "approved ground covers." White gardeners caught in violation of vegetation codes often enjoyed a green privilege. In Sugar Creek, Montana, a change in the law made Nathan Athans's garden illegal. Athans appealed the order to sod over his yard, and city officials granted that his garden was grandfathered in. City leaders threatened Julie Bass of Oak Park, Michigan, with a ninety-three-day jail sentence for her front yard garden. After press sympathetically covered the story, they dropped the charges. Martha Riley, in St. Louis, returned home one day in 1994 to find that city officials had mowed down her edible front garden. City officials admitted Riley's case was probably a mistake. White defendants found it possible to shift vague laws in their favor, but this was rarely the case for nonwhite homeowners. Tom Carroll and Hermine Ricketts, a mixed-race couple, had cultivated a front yard garden for seventeen years. The garden produced 80 percent of their food. Miami Shores Village Council passed a law banning front yard gardens and forced the couple to dig up their garden. Ketha Robbins in suburban Columbus, Ohio, sought to restore a forest to her backyard and landed in court.

As anti-gardening cases multiplied, a "Gardening While Black" hashtag appeared on Twitter. In a Detroit case, a Black man was harassed by two white women and then taken to court for gardening with children in a public park. In another incident, a man received a ticket for his front yard garden despite no restrictions being listed in the town codes. In Tulsa, months before a scheduled trial to discuss the legality of Denise Morrison's medicinal plant garden, city workers ripped it out.

In the Minneapolis suburb of Falcon Heights, city officials caught wind of Quentin Nguyen removing the sod from his front yard to plant a large garden as part of a neighborhood green initiative. They quickly imposed a one-year ban on front yard gardens (although the former city mayor had just such a garden). Nguyen, a garden activist, planned to use his two-acre front yard to grow vegetables to give away. But neighbors fretted that the garden was a security issue: "I am worried about normalizing the presence of many different people in front yards during potentially all hours of the day without any kind of restrictions put on access. . . . Many of these people were from outside of Falcon Heights which means that they are not necessarily familiar faces."

A neighbor and reported friend of Nguyen testified that the dispute was not "personal." It "was not racial." Nguyen, who came to the United States from Cambodia as a boy, responded with angry disappointment. He saw his garden as a venue for reciprocal relationships. "I wanted to pay back. Help the community. I feel targeted. I feel my fundamental liberties are invaded."

The demographics of communities with vegetable garden bans provide some context to these cases. During the decades of the 1990s to 2000s, racial segregation continued to decline in US municipalities, but at a strikingly slowed pace, while segregation by income increased. Scanning the national database of municipal regulations, my research assistants found that communities with vegetable bans had 30.4 percent fewer Black people, 28.5 percent fewer Asian people, 7.8 percent fewer Hispanic people, and 12.6 percent more white people than the population of their state. Those communities were also richer, with a 73.5 percent higher median income. This demographic profile supports what geographers call an "ecology of prestige" in which residents of higher-income communities attach greater importance to turf grass. Seeking a deep-green lawn, they apply more pesticides and fertilizer and influence their neighbors to do the same. The more homogeneous the social fabric of a community, geographers found, the less likely the neighborhood would have a biodiverse range of plants. In other words, the economic and racial diversity that American communities fail to cultivate is reflected in the plantscape. Mono-crop vegetation matches mono-crop demography.

That leads me to tell of another vegetable drama in the Windy City.

14 Hoop Dreams

IN 2013, NICOLE AND DAN VIRGIL lived in a lush, affluent suburb of Chicago. Dan had a good job. Nicole homeschooled their two kids.

Sifting through channels, they happened upon a documentary about industrial food production. "That's gross," Dan said. "Can't we just buy food that is food?"

Nicole asked around among her friends. "Where do you get real food?"

They stared at her, not comprehending her question.

"Whole Foods?"

Nicole drifted through the aisles of expensive organic food. Even pricey lettuce wilted and turned slimy after a few days in the fridge. She felt trapped, confined to the industrial food distribution network that girdled the globe.

I could try to grow a head of lettuce, she thought. It can't be that hard. She was not indentured to the corporate grocer. She was free. Free to grow a head of lettuce. Maybe more.

Nicole hesitated. Her houseplants had always died. Her parents were among the first Black families to move to Stamford, Connecticut, a corporate city. Her father broke the color line to run the city's music education program. Like many upwardly mobile families, her parents were eager to leave physical labor behind. Nicole did not recall anyone with a vegetable garden in her upscale suburb.

Nicole decided to plant her own garden. She and her husband, Dan, an engineer, don't do things by half-measures. They watched YouTube videos on gardening, checked out books from the library, and drew up plans. They built a raised bed and dug a wicking reservoir under it to store stormwater and drain the swampy, clay soils. They experimented with two plots. They dropped seeds directly into the spaded-up lawn and other seeds into a fertilized raised bed. Most seeds rotted in the clay soils of the lawn. Those that germinated did not thrive in the nutrient-poor earth, but the seeds in the raised bed sprang up in a few days and thrived, producing in coming months delicious vegetables of deep vibrant colors.

While in the Chicago area, I visited the Virgils in Elmhurst. The Virgils' front yard is neatly maintained with native flowers and turf grass. From the street, I could see no sign of a garden out back. In the backyard, Nicole gave me a tour of her garden made up of three long beds stuffed with produce. She and Dan had constructed frames from electrical pipes for the tomatoes and cucumbers to climb ten feet high. Enjoying the shade beneath, potatoes spread. The leaves of bush beans, spinach, chard, and kale fluttered in the breeze. Their plants had it all. Air, water, light, and soil. Nicole proudly showed me her cherry tomatoes grown from seeds she saved each year. Adapted to the microbes in her soil and their microclimate, the tomatoes were unusually large and a rich red color. She pulled vegetables from the garden, and we sat down to a delicious salad in their sunny kitchen.

In summer months, Nicole said, the garden provided 90 percent of the family's produce. Often, they had too much.

Nicole suggested the kids set up a vegetable stand as a homeschool project to sell their surplus. Nicole hoped her son would get better at doing math in his head. Before they started, Nicole and the kids made a trip to the Elmhurst director of zoning to make sure their vegetable stand was legal. It was. The city did not regulate children's businesses. The kids designed a logo, got signs printed, and set up a card table on Saturday mornings.

Neighbors asked, "Where did you get all this produce?" The kids showed them their backyard garden. Nicole recalls a woman snapping photos and calling her friend. "Did you know, lettuce grows from *dirt*?!"

The vegetable stand was a success. Each child made about $40 in spending money. Nicole's shy son learned how to subtract and talk to strangers.

Autumn comes swiftly to Chicagoland. The Virgils hated to stop gardening. On the web, Nicole noticed farmers in Maine extended the growing season with long plastic tunnels called hoop houses. You can buy hoop house kits for a couple of hundred dollars, but the Virgils are do-it-yourself people. Dan drew up plans for a wood frame connected with PVC pipes. He shored up the supports so the tunnel could withstand eighty-mile-per-hour winds and heavy snow loads. He carefully calculated the height and width of the tunnel to maximize passive solar heating inside. They located the hoop house in the middle of the backyard so it was not visible from the street.

The one thing the Virgils did not think about was the city's zoning board. Dan and Nicole had lived in Elmhurst for several decades. Elmhurst is a town of squat, white-trimmed, yellow-brick ranch houses placed in the center of spacious lots like iced pastries on a tray. Green lawns frame the houses. The lawns are largely unfenced, rolling along block after block, connecting one neighbor to another, a green communal thread. The Virgils saw neighbors build hockey rinks in their front yards and assemble trampolines and outdoor living rooms in their backyards. They figured the hoop house fell in the same category of a temporary recreational structure. They didn't count on one neighbor calling the city asking if the hoop house needed a permit.

One day they came home to find a Property Maintenance Violation Notice on their front door. The city required a permit for their "greenhouse." The Virgils stopped building. Dan went down to City Hall and explained their goal—to extend the season for a few months. They were not building a greenhouse. They'd take the hoop house down in the spring. He came away with the understanding that as long as the hoop house was temporary, it was OK, like the skating rinks and summer cabanas.

The Virgils loved the hoop house. Temperatures were as much as fifty degrees warmer inside. With snow all around, they worked indoors in short sleeves. The plastic frame functioned like Parisian farmers' hotbeds, extending the season. The family harvested fall crops in December and got a two-month jump on planting spring crops.

The next fall, frosty winds blowing, they put the hoop house up. Again, the neighbor phoned city hall, this time repeatedly. The neighbor was worried about how the hoop house looked, about possible flooding, about noise from the plastic cover flapping in the wind.

Another citation followed. The Virgils were confused. The City of Chicago allowed hoop houses. They could not find any rules banning hoop houses in the Elmhurst Municipal Code. Indeed, their citations listed no code violation at all. They asked what code or ordinance they had transgressed. City staff didn't know. They said they would get back to them on that. A letter arrived a few weeks later stating they had violated two codes, one about occupancy of a tent or temporary structure in a residential zone and the other from the permanent building code banning membrane structures altogether. That was confusing; how could the hoop house be both a temporary and a permanent structure?

Elmhurst officials told the Virgils they could get a variance to change the municipal code, but it would cost $6,500 for their $100 structure. The Virgils asked if they could work with city council members to write a new code. That would be fun, they thought, helping to draft new laws to make their community more sustainable. The kids would get a civics lesson. Two city council members offered to sponsor the bill, but the city manager would not entertain the idea. He told them to take the hoop house down.

The Virgils got a lawyer, had a hearing, and lost. They won on appeal but lost again when the city attorney appealed their victory.

The suggestion that she could not pursue a dream lit a fire under Nicole. Nicole is like that. When her high school music teacher told her she did not have the voice for opera, she trained even harder and became an opera singer.

Nicole called the local CBS news station. She talked to print reporters. Supportive neighbors formed a Facebook group. Nicole started a website, circulated a petition, contacted urban farms and nonprofits. She is articulate, funny, calm, intelligent, and informed to a granular degree. Nicole is a formidable force, but she and Dan were up against two powerful US institutions—the lawn and the suburb.

Unlike most parts of the world, Americans divide land by use—residential, commercial, agricultural, and industrial. Most suburban residential areas are zoned for single family homes, a category that excludes nearly everything on a lot but structures defined as "basic accessories," like garages and swimming pools. Like the rows of corn surrounding Chicago, Elmhurst is a mono-crop field with one product—bedroom communities dedicated to rest, relaxation, and consumption.

The Virgils' hoop house confused city leaders. At city council meetings, they asked, "What are you going to do with all that food?"

"Are you growing food to sell—a commercial venture? Raising food to live on, a subsistence farm?"

A city council member speculated improbably that an (unheated) hoop house could be used to grow marijuana in Chicago in winter.

Why, they were asked, didn't they just go to Whole Foods like everyone else? Residents in a middle-class community should be able to afford to buy their food.

Jane Jacobs wrote that suburban communities are utopias. Utopias are great, Jacobs noted, if you have no plans of your own. "As in all Utopias, the right to have plans of any significance belongs only to the planners in charge." But the Virgils had their own dreams—to produce nutritious, organic food sustainably and affordably. They were turning their suburban lot, zoned for residence and recreation, into a space for production, toiling with their hands like working-class people, instead of consuming, often ostentatiously, as American suburbanites are encouraged to do.

An Elmhurst city council member, Steve Dunn, instructed the Virgils, "We are a sustainable community but we are not a rural community." Zoning chairman Michael Honquest concurred: "This is a suburban setting, not an agricultural setting."

Nicole grew exasperated with these comments. "They tell us, 'We are not a farming community.' Well, OK, but I'm on two-tenths of an acre. I am not farming."

Like the Virgils, I was puzzled. The Virgils do not live in a homeowners association or gated community. They own their property. Why can't they do what they want on it within the parameters of the law? Despite repeated requests, no elected city official from Elmhurst would return my calls and emails. So, my confusion escalated. What harm did the hoop house cause?

It would have been easy enough for Elmhurst officials to investigate, but they didn't. Had they done so, they might have discovered that the windblown plastic did not roar like a train and there was no water runoff. On the contrary, the Virgils' trenched beds absorbed seven hundred gallons of rainwater each. Nicole claimed that when

they invited city council members to visit the hoop house, the elected officials declined. Instead, Elmhurst city officials poured thousands of dollars into judicial procedures and attorney fees to prosecute the Virgils. The case generated hundreds of hours of hearings and staff research. Why did they go to such expense and trouble?

If democratic procedure matters, then the case is even more puzzling. At city council meetings, a majority of attendees endorsed the Virgils' hoop house and "the right to garden." At council meetings over several years, less than a dozen people spoke against the hoop house while hundreds voiced support. Some supporters claimed racism was the underlying cause. Dan is white. Nicole is Black in a city that is 94 percent white.

"I am certainly conspicuous around here," Nicole noted. But in addition to race, she saw the controversy as circling class: "There's just this perception that it [the hoop house] is low brow and low class. That perception is next to impossible to chisel through."

Ari Bargil, a frequent litigator of municipal bans against growing vegetables, pointed to the general mandate for suburbanites to recreate, not produce. "What a lot of these restrictions are designed to do is prevent people from sustaining themselves." You can have a small raised bed, Bargil continued, "But they are saying that if you want to eat organic and nutritious food, you still have to go to Whole Foods like the rest of us."

Working with Bargil's Institute for Justice, Nicole campaigned to have her state representative draft a "Garden Act." She made several trips to the capitol in Springfield, lobbied legislators, and told her story over and over. Finally, six years of stubborn refusal to submit to what she considered arbitrary code enforcement paid off. The "Vegetable Garden Protection Act" passed in June 2021 gives Illinois residents the right to "cultivate vegetable gardens on their property or on the private property of another with the permission of the owner, in any county, municipality, or other political subdivision of this state." Illinois joined

Iowa, Florida, and Maryland in prohibiting homeowners associations and municipalities from requiring turf grass and restricting vegetables and native plants.

Nonetheless, Elmhurst's city mayor vowed to continue to battle against hoop houses in his city. The city council issued a ban on all temporary structures but finally permitted hoop houses after the Virgils spent a total of seven years fighting for them.

Buried deep in the hearts of Georgette Norman, Quentin Nyugen, and Dan and Nicole Virgil was a dream to create a small world in concert with plants—plants attuned to their yards and to them. Surely that can't be so dangerous? What would happen if a country simply embraced tiny gardens?

THE FIFTH TURN

The Dacha Empire

Superpower Self-Provisioning

ESTONIA IS COLD AND GRAY with rocky, clay soils—a tough place for growing food. That is why I chose to visit to learn how Soviets and post-Soviets used urban gardens. The dissolution of the Soviet Union carried out a kind of natural experiment in doing just what countries in the world now are aspiring to do: drastically reduce greenhouse gas emissions. From 1990 to 2010, the decades when Soviet manufacturing and industrial agriculture collapsed, greenhouse gas emissions fell by 35 percent in the former USSR, while they grew by 44 percent elsewhere. Estonian CO_2 emissions dropped 48 percent between 1990 and 2010. These savings in greenhouse gas emissions did not come at the cost of lost wealth. In the same decades, affluence grew by 54 percent. The Estonian miracle—radically cutting greenhouse emissions while enjoying economic growth—offers a glimpse of a successful transition away from fossil fuels. I wanted to know how they did that and how, in general, Soviets got by in the face of failed industrialization and agricultural projects over the course of twentieth-century history. I talked to gardeners to find out more.

The road to the Tallinn Airport in Estonia is straight, unswerving. To the left, concertina wire seals off the runways. To the right, trucks buzz in and out of warehouse distribution centers where the pavement turns to dirt, and a sign prohibits further passage. A notice dated 2018 announced that all tenants must vacate the surrounding territory. Standing there three years later, I watched caterpillars dig into the earth. The heavy machines sorted gray rock and gravel from black topsoil. Piles of discarded wood, windows, and fencing pointed to a seam of violence that had run its course.

When I walked past the "no trespassing" signs, the landscape shifted sharply from human order to botanical chaos. Under the airport's flight path, I brushed against a cluster of heavy, wine-dark berries—*Aronia Michurina*. I popped a handful damp with dew in my mouth: bitter, yet juicy, delicious. I noticed paddle-shaped *Aronia* leaves of a reddish tinge running in a straight course between hemlock, thistles, and honeysuckle. This was a clue. Plants don't grow in unswerving lines on their own. Botanical geometry points to human-drawn lines on a map.

Following the *Aronia*, I located what I sought: the ruins of a massive garden collective founded in the late 1980s. A decade ago, passengers on departing planes would have looked down to see thousands of people toiling in the earth in the Suur-Sõjamäe gardens.

During the perestroika era in the 1980s, the Soviet economy ran short of food. Employees of several large Tallinn factories received garden allotments on the city's periphery. In the economic crisis of the 1990s, factories laid off workers, and the garden collectives multiplied. They belted Tallinn and other Soviet cities with thick, green, edible corridors. But in 2015, city leaders announced an airport expansion plan. The gardens, which they considered an eyesore, would

Dacha garden in the semi-abandoned airport gardens, Tallinn, Estonia.

have to go. A few years later, bulldozers got to work shoveling aside fruit trees, berry bushes, potato beds, and lovingly constructed do-it-yourself cottages.

Despite the eviction notices, I found Valery Volkov, age eighty-four, still stubbornly inhabiting his plot and hut, called in Russian, a *dacha*. He and his girlfriend, Galina, seventy-seven, had taken over the allotments of departing neighbors. Valery and Galina had several large, snarling watchdogs, four goats, two sheep, a flock of geese, and more than a hundred chickens. They proudly showed me their fenced-off realm, a labyrinth containing a half dozen abandoned sheds, coops, and greenhouses with the last of the season's cucumbers, squash, and tomatoes. I pointed to zucchini ready to pick. Galina waved her hand, "I'm so tired of zucchini."

We stepped lightly around piles of scavenged windows that the couple were saving for renovations. Valery and Galina's world was one of remnants, the vestiges of a lost vegetable-powered Atlantis that was sinking into the swampy terrain from which it emerged.

The ruins of the garden collective revealed an undercurrent of Soviet history that doesn't fit the standard development narrative, in which rural migrants, like former peasants in Germany and England and former sharecroppers in the United States, land in cities and become urban. That narrative of farm-to-city and backwardness to industrialization does not make sense, especially in countries of the former Soviet Union. Instead, this history of modernizing urbanization played out in zigzag fashion with the shovel and hoe accompanying the assembly line, machine shop, and lab bench. Kitchen gardens frame Soviet history, from the turmoil of the Bolshevik Revolution to the end of the Soviet Union.

After the 1917 Bolshevik Revolution, urbanites fled large Soviet cities for villages and towns where they could maintain gardens. In the 1920s, they returned to cities and began cultivating yards and vacant lots. In 1928, Soviet leaders collectivized private farms, forcing peasants into large agricultural concerns that did not thrive. In 1932, during a man-made famine that killed seven million people, urbanites flocked to the margins of their cities and illegally dug and planted wherever they could. Rather than punish them, in 1933, the Stalin administration followed the gardeners' lead and granted citizens the right to grow food on public land. By practicing self-provisioning, Soviet citizens won the right to garden. And that saved lives.

A month after World War II began, Soviet leaders abdicated responsibility for feeding the civilian population. *Pravda* announced, "Each factory, each city is its own food base." Enterprises gave out state land. Employees farmed it. By 1944, sixteen million urbanites dug in parks, tree banks, and vacant lots. At the same time, every family in the Estonian city of Tartu received a plot to garden and each person was assigned a fruit tree and berry bush on city tree banks to harvest exclusively.

Estonia was forced to join the Soviet family of nations when the Red Army occupied the country during World War II. After the war, famine set in, and the number of urban allotment gardens multiplied, while urban Estonians relied on rural relatives for additional provisions. In 1949, Soviet leaders affirmed laws regulating the appropriation of public property for garden allotments. In the 1950s, Soviet Estonia industrialized, and thousands of migrants from Russia flowed in to take jobs. But that did not stop the spread of dacha allotments. Miners digging oil shale, chemical workers processing it, and technicians producing uranium yellowcake for nuclear bombs lived in segregated Russian-language colonies surrounded by state-planned garden collectives. By the mid-1950s, most potatoes in the USSR (60 percent) grew not in fields but in small plots on 7 percent of the country's arable land. A quarter of all vegetables and a third of eggs and milk came not from large collective farms but from tiny gardens. Gardens served the gardeners well. Researchers found that people who grew their own food ate better than those who did not.

The only dip in urban self-provisioning in Soviet history occurred when Nikita Khrushchev took power. Khrushchev, a former farmer, invested heavily in industrial agriculture and tried to put the brakes on small-plot gardeners by banning urban livestock, taxing garden produce, and moving people—including collective farmers—out of private homes with gardens into apartment buildings with no growing space. Taxed heavily, loads of people left their garden collectives. Food became scarcer, prices rose, and public support for Khrushchev eroded. The high price of staples led to food riots in Siberia in 1962, when soldiers opened fire, killing twenty-four people. Two years later, Khrushchev was forced to step down. His successor, Leonid Brezhnev, erased the prohibition on keeping animals, stopped taxing garden produce, and handed out yet more public land for self-provisioning. An upsurge in gardening followed in the years that most ex-Soviets see as the "Golden Age" of affluence.

In the small Estonian city of Paide, Tiuu Saarist's parents got a plot of six hundred square meters (0.14 acres) in their city's first garden cooperative. When I met up with Tiuu, she was more than happy to talk about her family garden. She pulled out photo albums devoted to the dacha. She showed me a picture of a field mired in five inches of standing water.

"This was what our garden looked like when we first saw it."

Directors of collective farms gave away land they considered worn out or unprofitable. In Estonia and most of northern Russia, that meant swampy, mineral-starved pastureland.

With shovels and scythes, Tiuu and her parents mowed, dug drainage canals, and pulled up woody undergrowth. "Hellish work," Tiuu remembered.

Starting with discarded land, gardeners had to regenerate exhausted soil. Some secured drums of nitrogen fertilizer from friends who worked on collective farms. But most scavenged resources around them. Ashes from the woodstoves, kitchen scraps, chicken manure, and human waste from the outhouse made up compost piles. Friends who worked on nearby collective farms dropped off cow manure.

Garden collectives pitched in at volunteer work sessions. Members of the Audit Committee would drop by to make sure gardeners complied with the rules. One important ordinance dictated that families exploit their plots only for food production. Untended vegetable beds, cottages larger than the mandated size, or dacha spaces used only for recreation (sunbathing and badminton) were verboten. Most people don't remember inspectors. They say they kept to the rules for fear of a neighbor sanctioning them.

On Friday afternoons in the 1960s and 1970s, urbanites, bags in hand, shouldering lumpy yellow-green backpacks would run for the bus, dash after a tram, and swing into commuter trains jammed with bodies, tools, bags of supplies, boxes of seedlings, crates of chicks. People heading to their dachas for fresh air first suffered the stale interiors of trains and buses. Like Tiuu's family, they spent the weekend building, digging, weeding, hands in the mud, feet in rubber boots. These were doctors, lawyers, teachers, professors, business leaders, party officials, salesclerks, cab drivers, and doormen—people considered middle class in the USSR. They enlisted their children in the work, which as adults they often remember as penal labor. Scan the peri-urban horizon on a spring weekend and you would see bodies bent over garden beds across nine time zones.

What kind of industrialized superpower had its citizens growing their own food? In the USSR, city people gave up weekend rest and recreation to weed, haul, and piss in cold, damp outhouses. A Canadian journalist described Soviet gardeners as reverting "from Industrial Man to Hunter-Gatherer." What a waste of labor power, two economists lamented, for valuable scientists to spend time digging potatoes, forfeiting hours at the lab bench.

No one considered the food Soviets grew or foraged to be alternative or sustainable. Their motivations were desperation; their actions a temporary coping strategy. By digging in the ground, Soviets lost the status of modern, urban consumers who, according to popular media, should have spent their time window-shopping and sitting in sidewalk cafés like Parisians in a Jean-Luc Godard film. Their humiliating neediness, the fact that Soviets gardened not by choice but for subsistence, made them failed subjects of a failed modernization project.

Tiuu's parents paid an architect for plans for a garden cottage. In 1967, city code inspectors approved blueprints for their tiny house, 215 square feet, a minimalist, Scandinavian design. By day, Estonian architects drew plans for massive, modernist housing developments. By night, they moonlighted to design playful, ecologically thoughtful, dacha houses for private clients.

The architects' plans included landscaping. Tiuu's drawings show six fruit trees alternating between berry bushes. The law bound gardeners to plant a minimum number of fruit trees. The ordinance was hardly necessary. Soviet urban farmers crammed a dozen fruit trees and two score of berry bushes onto their plots. Cooperative gardeners were instructed to plant fruiting hedges rather than fences to border their plot. A hardy berry bush, *Aronia melanocarpa*, grows everywhere you turn, populating many hedges.

A wildly successful plant, *Aronia* is also a technology. The plant breeder Ivan Michurin crossed an ornamental shrub, the North American chokecherry, with Alpine rowan to engineer a larger, tastier berry on a hardy stalk that grows in boggy, sandy soils in harsh northern climates. Michurin figured out how to carry out such long-distance hybridizations between two species and still produce plants that were not sterile (like mules). The shrub is excellent for dacha gardeners. *Aronia Michurina* propagates easily. The leaves repel mosquitoes. The bushes produces masses of large berries high in vitamin C. People made juice from them and dried them for teas with medicinal properties.

Michurin created thousands of new commercial hybrids for gardeners. At the All-Russian Research Institute of Genetics and Breeding of Fruit Plants in Leningrad, he and colleagues bred high-quality varieties of cherry, pear, plum, currant, gooseberry, grape, and apple, plants that thrived in northern climates, grew in short seasons, and resisted disease. During World War II, German Nazi leaders coveted this work. They created a special forces team that seized Michurin's plants and shipped them to Germany, where *Aronia Michurina* appeared on Berlin's tree banks to sustain malnourished Germans.

Despite the theft, Michurin's work continued. In the postwar decades, his followers, called *Michurintsy*, carried cuttings in buckets on trains and buses all over the USSR, as far as the Baltics. They planted their edible hybrid creations in parks and boulevards, and experimented in ways Michurin taught them to produce frost-resistant, self-propagating plants. In Estonia, the botanist Otto Kramer maintained a trial garden to breed plants for Baltic ecologies. Soviet gardening manuals showed gardeners how to make cuttings and graft fruiting branches onto hardy tree stock that could resist cold, damp, and pests in local microclimates. Across the country, Soviet plant breeders and amateur grafters pulled and stretched the pomological map, dragging fruits from France and Italy north to the short summers of the Soviet Union.

Dacha gardens became more important as Soviet collective farms failed. In the mid-1960s, faced with empty grocery shelves, Soviet leaders began importing grain from the United States, where agronomists were producing high-yield varieties of corn, wheat, and rice that could be cultivated with the help of lots of water and bountiful petroleum-fueled fertilizers and pesticides. For Soviet leaders, importing grain was embarrassing. A century before, the breadbasket of Russia and Ukraine had shipped wheat all over the world. Soviet scientists dreamed of a day when they would not need farms and fields at all. Microbiologists grew food in labs, devising caviar from petroleum. Chemists used imported palm oil to make artificial cheese, chocolate, and other synthetic foods. The leading nutritional chemist, Alexander Nesmeyanov, claimed that these new products would be purer than natural foodstuffs, free of bugs and germs, and, with sprayed-on vitamins, healthier.

In the same decades that Soviet scientists were inventing processed food, plant breeders were also designing hardy, fruit-bearing trees and shrubs for dacha growers. These two very different flight lines—scientific institutions dedicated to industrial food and others pursuing sustainable, small-scale growing—existed in two distinct economic realms. The Soviet planned economy plodded on alongside a gray-market economy where dacha gardeners experimented, traded, and shared produce from millions of tiny plots.

Gardeners had loads of fruit in the summer and fall, too much to eat at once. So, they typically dug fruit cellars ten feet into the ground where the earth maintains a steady temperature of 50°F. They stowed apples and root vegetables in the cellars. Everywhere I went in September, gardeners gave me bags of apples. They explained that they had summer apples, fall apples, and winter apples that are sweetest when picked after the first frost. Growing varieties with different harvest times meant that gardeners had a store that lasted until spring. Many apartments had fruit cellars in the basement where residents put up preserves and root vegetables. Carrots kept best in a box of sand. Apples went into a separate room because they release a chemical that rots potatoes and beets.

More delicate crops—cucumbers, tomatoes, mushrooms, squash, and peppers—they pickled. Pickling not only extends shelf life but brings about chemical changes that make food more digestible and make nutrients more accessible. Fermentation's conversion of carbohydrates to ethanol increases the caloric density of food—good for people in a deficit economy. Fermenting fruits and vegetables also elevates the free amino acid content—good for diets low in meat proteins. The health benefits of fermented food are enormous. Researchers are starting to understand the importance of fermentation for human evolution. Some speculate that it may have been as essential to early human brain development as the invention of fire.

Humans are not the only mammals that pickle to predigest their food. One day, gathering strawberries from the little edible forest Wesley and I cultivated in DC, I got exasperated about the rats that would take one bite out of a strawberry and move on to sample another, ruining half a row with their gnawing. I complained to Wesley.

"Didn't the rat mothers teach them to clean their plate? Why don't they just finish a strawberry instead of biting one and moving on?"

"Try one," Wesley replied. "The berries with rat bites are more delicious."

I grimaced at the thought, but, since Wesley is rarely wrong, I tasted it. He was right. The rat-bitten strawberry was sweeter and exploded on my tongue in a complex of tastes.

I learned why by reading an article by anthropologist Katie Amato. She and her collaborators found that apes take a bite of fruit, toss it on the ground, and return to eat it later after the bacteria in their saliva have fermented the fruit so that it is—like cucumbers in brine—more nutritious, palatable, and tasty. The smart, chubby rats surrounding our DC gardens were using the same fermenting technique as apes in the rainforest and Soviet gardeners in their root cellars.

Soviets may not have known the science behind their activities, but every gardener I talked to remarked that their food was healthier than any you could buy at the store. Soviet gardeners putting up vegetables and preserves saved themselves from a host of life-threatening health problems associated with vitamin deficiency during the winter, when there was no fresh produce in Soviet shops.

In 1950, a seventy-year-old Soviet could recall multiple deadly famines in her lifetime: 1891–92 (370,000–500,000 fatalities); 1921–22 (four million); 1932–33 (seven million); and 1947 (one to two million). After the postwar establishment of allotment gardens on a mass scale, the Soviet Union suffered no more historic hunger.

The Rabbit Eaters

SOVIET CITIES GREW A thickening green belt of allotment gardens around them in the same decades that Americans spent millions cultivating turf grass. Western observers classified dacha garden cooperatives as primitive spaces slated to disappear in large part because they lacked urban infrastructure—sewers, water, electricity, garbage collection—and had few roads or paved surfaces. Gardeners rode buses and trains, then walked or biked to their plots. Footpaths connected garden to garden. Gardeners drew water from wells or gathered it in rain barrels. They relieved themselves in outhouses, sprinkling sawdust or peat on top to dampen the smell. They gathered sticks and logs to burn in Finnish stoves. As more families built tiny houses on two-tenths of an acre, density increased.

Without sanitary infrastructure, dacha communities could have turned into slums teaming with bacterial pathogens, with gullies washing earth into waterways and mounting piles of garbage attracting rodents. In other spheres of the economy, Soviets were horrible stewards. They had trouble delivering food to markets, so produce piled up and perished. They struggled with quality control and supply chains, so mountains of parts, tools, and construction materials were left out to rot or rust. Meanwhile, radiation, heavy metals, chemical toxins, and nitrates poisoned the earth, water, air, and food. But in the dacha universe, a very different picture takes shape. Building codes and regulations directed gardeners in constructing low-tech sanitation facilities. Plans instructed how to build compost bins and dry toilets lined with clay and rocks to filter pathogens. Regulations mandated that a diverse mix of perennial plants, which hold earth, absorb water, sequester carbon, and feed birds and bees, surround every plot. Codes dictated that cottages could not be located near bodies of water and could be no larger than 270 square feet, about the size of today's tiny houses. These regulations are now seen as best practices for green architecture. If a Soviet garden collective were a contemporary suburban development, the planners would, like the architects in DC, win awards for sustainability.

Most people constructed their garden cottages themselves with hand tools, but building materials were hard to come by. When windstorms blew and took out swaths of forest, people had the right to harvest the windfall for lumber. Gardeners used other wastes too. They scavenged unused building materials at construction sites and made deals with friends who had access to excess goods.

Speaking of these exchanges between friends, dacha owner Mart Pungo told me, "You wouldn't pay a person with money. That could get you arrested. You just gave them a bottle of vodka, and he was happy to share with you our socialist wealth! If he got caught, he wasn't a criminal, just a drunk."

Little went to waste. Piret Valk described how her grandmother used plastic bags until they tore. Then she would cut them into strips to weave into mats. Pungo managed a restaurant. Nonreturnable bottles piled up in the alley. He took them home to build the exterior walls of his dacha (think stained glass) and a vaulted greenhouse of bottles. In the city of Narva in eastern Estonia, Pungo's naive architecture is so loved that tour buses regularly stop to see his creations. Dacha builders scavenged around dumps, riverbanks, and urban edges, giving a second life to discarded materials. They saw themselves as cleaning up the city, sanitizing the urban ecosystem.

As no garden store existed in Paide, gardeners took cuttings of apple and plum trees and berry cultivars from a nearby collective farm and an abandoned manor house. They grafted branches of trees known to produce good fruit onto hardy rootstock. Tiuu's parents traded seeds with other gardeners in the collective. This sharing economy is still visible today. Most garden beds have a similar retinue of vegetation—fiery-orange calendula, creeping pink and red roses, squat marigolds, spiky crowns of dill, green carrot fronds, red beet stalks, and delicate white potato blossoms. These same plants repeat from dacha to dacha across garden cooperatives. The thrift, self-reliance, and material autonomy of dacha gardeners show a care and sustainability rarely associated with Soviet history.

Gardeners in a collective worked together, while reportedly they had little to do with people in neighboring collectives. In Estonia, there were Russian allotments and Estonian ones. People also sorted themselves within dacha associations. Henn Sokk showed me around.

"This here was a cooperative of municipal workers. Over there was a factory cooperative. Down there were two guys released from the gulag. They formed their own cooperative."

Gardeners had access to chemical fertilizers, pesticides, and herbicides from local farms, and they used them. But most gardeners surveyed by a team of sociologists reported that they tried not to use chemical treatments. "If a cabbage leaf is bitten by an insect, it does not matter because this insect is useful for the balance of nature. . . . So

I do not spray my plot," a woman in Russia told the researchers. An Estonian gardener said, "If I use chemicals, then it is not food anymore." Tuuli Reinso explained to me, "I was taught to observe the plants, what they were doing, what they need. Rhubarb thrives next to the well. It needs a lot of water. Leafy greens are resilient and can grow in windy spots. Cucumbers are more sensitive." The small scale of the gardens and diverse interplanting made managing pests an easier task than keeping susceptible crops healthy in mono-cropped fields.

Tiuu said that once their garden got going in Paide, her family did not go to the store for food. They consumed fresh produce, but also created food to replace what was missing in Soviet shops. They processed beetroots into sugar. Rowan berries went into bread as a substitute for hard-to-find raisins. They had chickens for eggs and meat, and hundreds of rapidly reproducing rabbits for meat and fur. Tiuu grimaced: "I can't eat rabbit meat now. I had so much of it."

As the number of Paide garden collectives multiplied from the first dozen to more than a thousand allotments, residents grew more food than they could eat in a year. They gave sacks of potatoes, beets, and cucumbers to family members. They traded tomatoes that poured out of the plastic-walled greenhouses. Still, they had too much produce. Tiuu's parents took their extra vegetables, eggs, and rabbit fur to the town hall. Municipal authorities bought it, and the goods appeared in local shops. But soon, the garden bounty overflowed the few stores in the small town. City leaders began to sell gardeners' produce to other towns. They made the most lucrative deal with traders in Leningrad, exchanging produce for sought-after canning supplies, fabric, and rare translations of foreign novels. In the swap of carrots for Camus, gardens enriched consumer and cultural life.

Paide is typical of small-town USSR. By the end of the 1950s about one-third of all agricultural production in the Soviet Union came from private agriculture, not collective farms. Surveys show that a typical urban garden produced at least 440 pounds of fruits and berries and 550 pounds of vegetables, enough to supply a family of four with produce for a year.

While smallholding plots were extremely productive, Soviet industrial agriculture continued to wither. In the early 1970s, Soviet planners noticed a deficit between the agricultural targets of collective farms and actual production. In 1977, the crisis in Soviet agriculture grew desperate. The productivity of agriculture remained flat while investments in farming grew. Faced with food shortages, Soviet leaders passed new laws allocating yet more public land to private gardens and providing loans for dacha construction and planting. The share of Soviet agricultural goods that came from private smallholders increased even more.

In 1981, when a new popular magazine, *Kitchen Garden* (*Priusadebnoe khoziaistvo*), hit newsstands, Premier Leonid Brezhnev used the country's biggest political event, the Congress of the Communist Party, to underline the importance of tiny gardens on every urban horizon: "Experience shows that such holdings [garden allotments] can be an important additional source in the supply of meat, milk, and other produce. Individually owned vegetable and fruit gardens, poultry and cattle are part of our common wealth." By that time, half of all Soviet households, 46.6 million families, were members of a garden collective, three-quarters of them in cities. Far more people joined a garden collective than joined the Communist Party. Garden associations were the most enduring popular movement in Soviet history.

Gardens were popular for lots of reasons beyond food provisioning. During summer school vacations, city children went to their grandparents at the dacha, where they had clean air, space to play, and fresh food. Pensioners made extra income selling garden produce. When Soviets needed a pile of cash to pay for a car or apartment, they turned, not to doctors or professionals in the family, but to unemployed grandmothers who had built up large bank accounts selling one bunch of parsley at a time.

Gardens also helped Soviets find their place in a large and repressive empire. People moved around a lot in the USSR, uprooted for job assignments, for service in the military, or because of arrest and deportation. Anu Printsman's parents came from southern Estonia to settle in Kohtla-Järve, a grim mining town in eastern Estonia where her father mined oil shale, a low-grade coal. Her mother, a chemist, had a job at a chemical plant that combined oil shale from the ground and nitrogen from the air to produce ammonia fertilizer for industrial agriculture. Long plumes of black smog flowed from the plant's smokestacks, under which Anu's family had an allotment garden located over a shuttered mine. Sandwiched between fossil fuels in the sky and those underground, they gardened. Anu's relatives carried seedlings north on their visits to Kohtla-Järve. They used cultivars to stitch together seams of the uprooted family.

Anu listed the plants. "We have a good gooseberry that originates from Mom's birthplace. Our apple trees were grafted from Grandmother's garden. Tomatoes we called after my dad's mother."

While Estonians could draw on extended families and childhood friends to fill in the gaps of the patchy Soviet consumer network, migrants to Estonia from Russia, like Svetlana Trofimova, had to make do on their own. Svetlana grew close to her neighbors in her garden collective near the Tallinn Airport. They kept on eye on each other's children. Neighbors chatted over the hedge, shared harvests, and pitched in to help with construction projects. Gardeners often spoke about the desire to "have their hands in the ground." The best "sweet" soils, they said, are dark in color, easily worked, alive with insects and worms. In contrast to farmers who use large machines, gardeners work with bare hands. Digging in vegetable and flower beds, growers exchange microbes with their soils. Giving away homegrown food, gardeners also share their microbiome among friends. As with migrants to DC, Soviet migrants used such biological exchanges to help adapt to their new place. Soviet science emphasized nurture over nature, environmental influences over genetics. Recent insights in microbiology and epigenetics justify this focus. Garden collectives served as both cultural and biological mixers.

Home has many connotations. For many people living in the USSR in the late 1980s, home meant crisis and the collapse of the Soviet economy and state. In 1987, with food in shops growing increasingly scarce, Premier Mikhail Gorbachev turned to gardens as had Soviet leaders before him. He encouraged citizens to feed themselves. His administration passed regulations making it easier to acquire a garden allotment. Garden associations spread across hundreds of thousands of acres of farmland that had been abandoned by struggling collective farms.

In the late 1980s, state enterprises and institutes ran out of money to pay employees. Instead of wages, employees would get an occasional food package or goods made in the factory, which employees could exchange for what they needed. This barter system was cumbersome. I lived in Moscow during this grim time. In 1990, prices, which had long been stable, climbed by 200 percent and kept going. A person's lifetime savings dried up in one trip to the market. My landlady told me food was so expensive that she lost weight. She showed me before and after photographs, preening a bit to display her newly trim figure. She was a relentlessly positive person.

"This is better. We didn't need to be eating all those sausages."

But things got worse. In 1992, after the Soviet Union dissolved, food grew yet more scarce and inflation spiraled out of control. As the days darkened in autumn, Muscovites recalled the mass famines their parents and grandparents endured in the first decades of the communist state, and stashed sacks of potatoes and beets under their beds.

In response, millions more Soviets took up hoes. Zinaida Vasilyeva's grandfather had advanced in his life from a village boy to a military engineer and international consultant. He and his wife, also a villager, loved to spend their holidays, not at a dacha digging in mud, but under a parasol on the beach in Crimea. In the early 1990s, he was forced to go back to his native village to ask for a garden plot. For him, returning to the village to farm was a humiliating reversal of his social mobility. Even though the village was in the Russian Northwest in a "zone of risky agriculture," the allotment grew to great importance in supplying the extended family with hardy vegetables and, in warm years, strawberries and tomatoes.

In Tallinn, factory managers, unable to pay their workers, raced to create new massive garden complexes to accommodate employees who needed to self-provision. Employees of the Dvigatel factory expanded the airport gardens. Workers at a chemical plant plotted out another massive garden association north of the city in Maardu. The appearance of these new garden collectives reflected the disintegration of the state and economy. Architects no longer designed cottages, and city officials no longer approved plans and enforced code. People built whatever they could with whatever they could find. Instead of the sleek Scandinavian designs, late Soviet dacha houses were like depression-era huts East of the River in DC, patched together with found materials. Worried about rising rates of crime, gardeners constructed tall fences and invested in watchdogs, chains, and locks.

With medical services collapsing and pharmacies empty, gardeners read up on herbal remedies. They described how calendula was good for pain and cramping. Boil it, mix it with goose fat, and you have an ointment for burns and wounds. Lady's mantle was recommended for a throat infection. Ivan Chai was said to help with spring fatigue and gave nutrients and strength. It was also said to fortify male potency.

At a time when most wage earners were out of work or unpaid, gardens became a crucial source of income. In the 1990s, Olena Palko's father had a conflict that made his job unbearable. He took to drinking. Her mother convinced her husband to quit his job; they would find some other way. At that time, the government of independent Ukraine was giving away collective farmland to anyone who wanted it. Olena's parents acquired a quarter-acre plot just outside their small city, Shepitivka, in western Ukraine. Olena's father dove into gardening. He planted long rows of vegetables and set down an orchard. He read gardening magazines and visited agricultural exhibitions. He carried out experiments with new seeds and varieties that he ordered by post. The whole family was expected to pitch in on her father's and grandparents' plots. All weekend and after school, Olena, her parents, and her grandparents would be in the garden, bent over rows. To pass the time while working, Olena's grandmother would retell the plots of her favorite novels. For transportation, they had just two bicycles for five people. The bike carted water from the well to the field. Her parents balanced sacks of potatoes on it to carry them home. For Olena, the garden was a symbol of their poverty and desperation, but everyone else in town was doing the same.

"We bought oil, flour, salt, and sometimes meat, that's it. The rest we ate out of our garden. We had little else. If we had lots of zucchinis, that's what we ate. If the cabbages didn't thrive, we didn't have cabbage."

They spent days pickling and canning. Olena had the humiliating job of selling produce from a box at the local market and bus stop.

"We got just pennies for our berries and greens. Nothing. But we were so poor, we needed it."

Shame is a sentiment that runs deep in Soviet culture. As media from abroad appeared on TV screens, Soviets learned that they did not measure up to the gleaming, technicolor standard of life in the West.

In 1991, Svetlana Trofimova lost her job at the nuclear parts division of the Dvigatel plant. Independent Estonia had no nuclear weapons and no plans for them.

"My husband and I both worked there. Suddenly we were out of work. No money coming in at all. That is when the garden saved us. We had two boys to feed. We grew everything we could. Potatoes, onions, greens, squash, tomatoes, cucumbers. We had fruit, black currants, red currants, aronia, and raspberries. I sold garden produce from the sidewalk. I wasn't too proud to do that. We needed the cash of course."

It was hard to think of Svetlana and teenage Olena standing in a gray drizzle over small boxes of spindly parsley, dill, and onions. How many women like that did I walk by in my passages through the cities of the former Soviet Union? They would be there late in the evenings in the rain or snow and early in the morning. Perched in front of subway entrances, they served as sentries to the post-Soviet collapse.

Svetlana counted on her hand the years that her family relied on her garden to scrape by. She ran out of fingers. For a full dozen years, the former nuclear technician dug for subsistence.

Across the former USSR, the number of garden collectives nearly doubled in the 1990s, in Russia alone from thirteen million to twenty-two million. In the same decade, production in the highly subsidized sector of industrial agriculture dropped by nearly half. Large state-run enterprises farmed one hundred million hectares, but by 1999, they produced less than all the family plots in the country did on a total of ten million hectares. A Lithuanian tractor driver remarked on the great difference in productivity between farm and garden.

"Everything grew by itself! . . . I swear I grew ten times more cabbages on my tiny lot without any technology than the [state] farm ever did using all the chemicals."

As Soviet industrial agriculture failed, urban gardeners with burlap sacks on commuter trains displaced farmers mounting tractor combines. The food grown by engineers, diplomats, professors, janitors, and assembly-line workers made a critical difference in preventing the feared post-Soviet famine. Gardeners' share of agricultural production doubled in the 1990s from 26 percent to 52 percent. By 1996, smallhold gardeners grew 91.9 percent of Russia's potatoes, a key staple. That is a staggering fact, given that community gardeners occupied 1.5 percent of the arable land. Western economists wrote up a prescription for "shock therapy" to stimulate a transition from Soviet socialism to

market capitalism, but the remedy failed to supply enough food for 287 million citizens. Instead, the gray spaces of garden collectives, neither socialist nor capitalist, put food on people's tables.

And the Soviet Union was not alone. A similar tide of urbanites flocked to urban gardens in Cuba in the 1990s in response to the end of Soviet subsidies and a new round of US-imposed embargoes. By 2002, small-plot gardeners grew 60 percent of all fresh produce consumed in Cuba, a country that had long imported most staples while growing sugar and tobacco for export. Homegrown fresh fruits and vegetables enhanced Cuban food security and overall nutrition. In Eastern Europe, between 35 percent and 60 percent of the population grew some or all of their food. Few pundits noticed the miracle that millions of tiny gardens fended off hunger in relatively peaceful transitions from socialism.

17 Flowers

ONE OF THE MOST commercially successful state farms in Soviet Estonia was located on the suburban edge of Tallinn. The Pirita Sovkhoz specialized in hothouse flowers. A long indoor street connected several acres of parking-lot-sized greenhouses, each of which cultivated a different flower. Heating pipes ran through the glass ceilings and heavy cast-iron potting tables gridded the floor. The complex was massive. An employee could walk indoors for the length of five and a half football fields to reach the farm's club and bar. The lights of this crystal palace burned all night, illuminating the high cliffs above the Baltic Sea like a Broadway marquee.

The greenhouse complex had its own gas-heating plant to keep the flowers warm in the winter. A pond fed the boiler and provided water for irrigation. The Pirita flower farm burned the vegetal past to fulfill a desire for exotic, botanical treasures that, thanks to fossil fuels, could grow anywhere, at any time, in any weather.

The carbon sent into the atmosphere from the Pirita flower farm floated upward to mingle with the carbon released from coal furnaces in Karl Marx's time, when he famously penned "all that is solid melts into the air," describing his daily swim through the carbon-packed smog of nineteenth-century London. Two centuries of carbon released into the atmosphere works like the glass panes at the Pirita flower farm, snaring gases and heat, turning the planet into a greenhouse much larger than any Soviet megaproject.

Today only a few sections of the Pirita greenhouse are still in operation. The farm was privatized years ago. Several employees purchased individual greenhouses and kept growing, but most of the flower complex slowly sank earthward to rest in beds of shattered glass. I visited the farm with my friend, Linda Kaljundi. We came across an older couple, former employees of the state farm. Their hothouse was crawling with grape, cucumber, squash, and tomato vines. They were busy, but paused to talk to us while Linda bought tomatoes and interpreted from Estonian.

The farmer showed us an oil shale by-product, a thick black tar that she mixed with water and sprayed on grapes to kill a mildew fungus spreading over the leaves. She explained that in Soviet times, they fought blight and insects in the hothouse mono-crop environment with a full arsenal of pesticides. They applied poisons, she said, without respirators or protective clothing. Several of her coworkers tasked with spraying got cancers and died at age thirty or so. She attributed their deaths to the toxins.

Hothouse flowers were grown with chemical nitrates, heated with petroleum products, and defended against pests with chemical by-products of the petroleum industry. Combustion engines drove the blossoms around the USSR to flower shops, hospital kiosks, and sanatoria. Once purchased, the bouquets traveled from hand to hand and into oncology wards where patients received them. The flowers and cancer patients shared a common feature. The petroleum products saturating the vascular structure of the blossoms also circulated in the bodies of patients treated with chemotherapy. Both the flowers and the patients were a product of petro-modernity.

But that is just one part of a larger story. Down the street, Pirita farmworkers lived in freestanding houses with large yards where after hours they farmed in a different way. They built their own, much smaller greenhouses that warmed passively in the sun. Around their houses spread garden beds with the usual mix of fruit trees and berry bushes. At the flower farm, Pirita farmers mined topsoils of nutrients. At home, they built soils from the ground up. On their own time, hothouse farmers became gardeners. As they walked home from their day jobs, they shifted from petro-modernity to retro-modernity.

What does retro-modernity look like? Soviet gardeners closed the cycles of extraction that impoverished soils and led to repression and mass famines in tsarist Russia and the Stalinist USSR. And they had help. Postwar Soviet law and cultural institutions aided in shaping garden communities. While officials in the United States passed hundreds of city ordinances mandating the growing of turf grass, Soviet regulations supported garden collectives and protected waterways, soils, and public health. The state invested billions into land grants, education, and botanical services. State-sponsored TV programs, magazines, books, and courses showed gardeners how to make compost; when to plant, harvest, and preserve; and how to prune fruit trees and deal with predators. State-owned nurseries developed seeds and seedlings of hardy varieties expressly for dacha farmers.

Around the world, small-scale farmers produce an estimated one-third of the world's food. Most often these farmers are envisioned to be located in the global south, yet Russia and countries of the former socialist block have the highest number of small landholders in the world. Soviet kitchen gardens were one of the most economically suc-

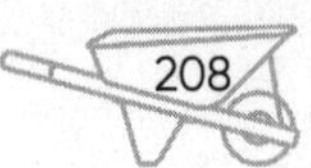

cessful and sustainable production sites in the Soviet polity. The Soviet gardeners' state built a whole apparatus directed at what became the world's largest urban farming endeavor in human history.

Unfortunately, Soviet laws and public land (commons) that protected gardens departed on the same train that replaced the socialist state. Bulldozers razed many thousand gardens that were built chaotically at the end of the Soviet Union, such as those near the Tallinn Airport. The gardeners, mostly Russian-speakers who did not have Estonian citizenship, did not dare to raise their voices in protest.

When I asked Svetlana what it had meant to her, several years after her garden community had been razed, she just blinked, holding back tears. She later told me how, when her dacha plot was destroyed, she cried and cried for months. "It was painful, painful! For years I grieved that loss."

In Estonia, new laws in the 2000s enabled people to privatize the public land under their dachas. In desirable areas, people built larger houses and supplanted potatoes and berry bushes with turf grass and trampolines. Steadily the grand garden collectives surrounding Estonian towns and cities transformed into suburbs and exurbs.

As the dacha green belt dissolves, new dacha suburbs contribute to sprawl and Estonians drive more than all other Europeans. The country relies on low-grade coal for energy and imports more food than ever before. As a consequence, greenhouse gas emissions in Estonia are among the highest in the European Union. In 2023, Tallinn won the title of European Green Capital. Unfortunately, that award came a few decades too late.

THE SIXTH TURN

Mansfield, Ohio

18
Terrible Farmers

IT HAD ALL THE MAKINGS of a reality TV show. Ten contestants competing for the chance to win their own business, an urban farm in Mansfield, Ohio. All they had to do was grow a certain number of cucumbers, kale, and tomatoes for three years. Who would buckle under the strain of farm labor? Who would outproduce the rest? You might bet on Tim, who works for the Farm Bureau, or Amanda, a master gardener, or Vinnie, known for his green thumb, or maybe Walt, a former football player with a lot of business acumen.

But it was not reality TV. No cameras were rolling, there was no rivalry, nor even a measure of the Jeffersonian vision of the bootstrap American farmer in this scenario. None of the contestants expressed a lifelong passion to till the ground. Instead, the ten people were linked together because they had passed a training course and formed a co-op, or a "gro-op," as they called it. Their goal was not only to grow food but to plant opportunities in a rustbelt city where working-class jobs dried up in the 1970s.

Mansfield used to manufacture brass streetcar fittings, farm implements, and other goods that were purchased around the world. Then the factories closed, taking union jobs with them. Now, highways tentacle Mansfield, stretching like gooey taffy to steer cars away from the city on seven hills (like Rome, residents will tell you) to shopping malls and prison complexes on the outskirts.

Mansfield inspires sentence fragments. Handsome, crumbling Victorian mansions. Engine parts scattered under the chassis of an Oldsmobile. Razor-wire outlines of correctional facilities, lights blazing against the night sky. Teen parents, a toddler on the hip and baby in a stroller, heading into the Dollar Store. In Mansfield's working-class North End, acres of boarded-up factories and empty lots roll along like a silent home movie.

The contestants who sought to win a spot in the gro-op each had their own backstory.

Charlotte, who runs the school garden, raised many kids over the course of relationships with four different men. The first turned out to be a drinker. The second beat her. The third took a liking to her daughter. The fourth decided he would no longer work. Each time a husband failed her, Charlotte gathered up all children from the union—his and hers—and left.

Amanda told me, "For the first decade we lived in the North End, our intersection was the major corner for drug traffic in Mansfield. One day a law enforcement official knocked on our door and asked if they could use our house for stakeouts. "I asked him why he didn't worry that we were one of the drug dealers, and he said, 'Because you mow your lawn and plant flowers.'"

Walt, with a master's in business, was going places until one night three white men jumped him in a bar parking lot. The security video footage shows Walt, arms out, backing away from the fight and out of the camera's frame, where his assailants let loose. The men pummeled Walt, smashing his head into the curb. The police officer who responded turned out to be the brother of two of the attackers. The officer handcuffed Walt, his skull cracked, and let his brothers go free. Facing a jury with no witnesses, Walt took a plea deal for a violent felony with no prison time. As a convicted felon in the United States, you can become president, but you can't vote, rent an Airbnb, or get a business loan, and it is very difficult to find a job that pays a living wage.

When I first met Vinnie pruning tomatoes in his hoop house, his son had been dead of an overdose for just three weeks. His wife had a skeletal appearance from four years battling cancer. Vinnie told me he had worked a lot of jobs over the years. He listed them: "US navy submarine sailor, chimney sweep, busboy, Long John Silvers, pizza delivery, construction, fire extinguisher salesman, baker, window washer, painter, janitor, gentlemen's club manager, landscaper. . . . I've made a lot of people a lot of money, but we have always struggled with poverty."

In the vacant lots of Mansfield, Charlotte, Walt, Vinnie, Amanda, and the others bury their sorrows in long rows of raised beds, hoping to harvest something better. Vinnie dreams of becoming a philanthropist, giving away money to people struggling to pay bills. Walt seeks to build an empire of healthy food and healthy jobs in Mansfield, where he grew up. Amanda wants to replace the drug dealers on her street corner with fresh vegetables. Charlotte wants to teach kids about plants so the children might know from whence their future springs.

The Richland Gro-op established the main part of the farm on a twelve-acre lot on Bowman Street, formerly the grounds of the Gorman-Rupp Pump Factory. They constructed hoop houses from metal rods and sheet plastic. In them, they pieced together raised beds. A supplier dumped piles of a product he advertised as compost. Vinnie shook his head.

"I told them that stuff was no good."

Ken, a soil scientist from Ohio State Mansfield, who was advising the group, grimaced.

"No one asked *us* about the soil."

Seeds didn't hold in the grainy mix of leaf mulch and sand. Water ran through it. That is, if they had water. After they had planted seedlings, someone noticed they had no water supply on site and no budget to pipe it in.

Walt described the root of their troubles the first year. "The co-op's biggest problem is we don't know how to farm. We are terrible farmers!"

Most farmers learn the trade from their parents. None of the gro-op members had agricultural backgrounds. Only a few had ever even gardened before signing up. Tim laughed: "I work at the Farm Bureau, but I don't know how to grow anything." Walt was inspired by an uncle who had a backyard vegetable plot. Marqua had long grown vegetables for her family. She had beds of rich blue-green kale, but she admitted she didn't much like kale. "I have an unhealthy addiction to cookies, candy, and bourbon."

Oddly enough, this group of unlikely farmers was the brainchild of a history professor at Ohio State Mansfield. Ken "Kip" Curtis taught in Florida before he moved to Ohio. There, his children went to a local school where a number of the kids qualified for free school lunches. Kip saw that the food didn't look appetizing or healthy. There they were in sunny Florida. He started a school garden that was wildly successful, spawned seven others, and spun into a food literacy program. The scale was small, but the impact felt powerful.

When Kip moved to Ohio, he had a vision of something bigger. Kip grasped that Mansfield lacked access to healthy food as well as jobs, job training, and opportunities. What if they could turn some of the many vacant lots in Mansfield into urban farms? Would that help? A New York University study found that two-thirds of urban farmers make no more than $10,000 a year. Kip speculated that while it's true that one farm can't generate much income, ten urban farms working together would amount to an agricultural concern large enough to attract buyers so that together the farmers could net, he calculated, about $30,000 annually. Kip set up meetings and formed an alliance with Deanna West-Torrence, the founding director of the North End Community Improvement Collaborative (NECIC), a neighborhood nonprofit dedicated to reviving the poorest part of Mansfield.

Deanna and Kip hit it off. Deanna grew up in the North End. Her great-grandparents had arrived in the 1920s in the great migration.

"It was a wonderful, wonderful neighborhood, but the newspapers portrayed it as a mess."

At the time, in the early 2000s, the North End was suffering from massive divestment. The city had closed all six schools in the area and, consequently, cut policing. Deanna ran for city council to turn that tide around. At the time, she worked for a community health center. Later, she left the council and ran for the school board. Working in government, health, and education, she pieced together how problems in one sector amplified those in another. She wondered how it would be possible to fix so much at once.

And then something miraculous happened. The family-run Rupp Foundation went looking for young community leaders to invest in and contacted Deanna. They took her to dinner and asked her what she would do with a large grant. She said she would start a center that would focus on what was needed in the North End—education, health care, jobs, and youth development. The foundation gave her grants and eventually, a large endowment that has supplied the funding for her organization. They also provided Deanna, who had a high school education, with advisors and mentors to help her figure out how to most effectively run a nonprofit.

When Kip came to her proposing to scale up gardens to urban farms, Deanna was all in. "We didn't know we were a food desert," Deanna told me over the phone, "until we announced a small grants program, and a flood of requests to start community gardens came in." At that point, Deanna realized that her neighbors acutely felt their inability to access fresh food.

The residents of Mansfield's North End did not need more charity, Kip and Deanna agreed, but they did need a vehicle for people to get on their feet again. However, most urban farming ventures emerging from community activism did not have staying power because they relied on grants, and funding appetites change over time. Kip's vision involved setting up opportunities for entrepreneurs to produce healthy food and also create jobs to draw the city out of cycles of poverty, abandoned houses, declining tax revenue, and failing infrastructure. Mansfield's longstanding economic decline worked in the gro-op's favor. The North End had plenty of abandoned land, and it qualified for state and federal grants targeting low-income communities.

Kip enlisted educators and soil scientists from Ohio State University extension. He got the use of an old tennis court on his campus to set up hoop houses to hold training workshops for gro-op members. He furiously wrote grants. He won $2 million to create a new cooperative model for urban agriculture. His success in organizing and winning grants makes him a rock star among history professors, who are traditionally trained to teach, research, and write, not start businesses or partner with community nonprofits.

Forming a cooperative might seem like a return to the naive idealism of another era, but the most successful farmers in the United States today are those in cooperatives. Ohio has the largest populations of Amish and Mennonites, who make up the fastest-growing sector of new farmers in the United States. The Anabaptist orders live and work collectively, pooling labor, equipment, and finances. They tend to be market-shy, growing and preserving much of their food in communal kitchens, sewing their own clothes, and reinvesting profits back into the community. They raise a wide range of crops and livestock, requiring fewer acres than conventional farmers. As a community of

Anabaptist farmers grows, several families will break off to form a new colony. Farmers in Ohio outside Anabaptist cooperatives have trouble competing. They grouse about the Amish who pull up at a land auction in horse-drawn buggies and slap down cash for high-priced real estate. After the Richland Gro-op got going, it would partner with these Amish farmers.

Having established the Richland Gro-op, Kip handed over his leadership role to the co-op members. Their first job was to find customers. Walt organized logistics and managed the organization's urban farm. The group hired a marketer to find buyers for the vegetables that co-op members would grow. The marketer handed seed catalogs to potential customers, restaurants, and grocery stores, and told them the gro-op would grow whatever they wanted. The more exotic, the better. Fifty pounds of Hokkaido radishes and purple carrots? Sure. Microgreens and heirloom peppers? No problem.

A few local restaurants and a grocery store signed on to make regular purchases. With buyers lined up, the gro-op drew up a growing plan. Each member committed to producing so many tomatoes, so many pounds of kale, cucumbers, and the rest. Grants provided every farmer with two hoop houses and two large, outdoor raised beds, amounting to a tenth of an acre. Some members had their farms next to their homes. Four farmers worked at the large vacant lot in the Northeast end of Mansfield.

The first spring, growers laid down plant starts in the hoop houses, but with no water, the seedlings would die. Walt and his girlfriend Nakeia had to drive over to the community center to fill a large plastic water tank in the bed of Walt's pickup. Pushing and juggling the heavy container, they irrigated the spring starts. For a month, they spent long hours on the simple act of watering until the gro-op got a water line and installed a drip irrigation system.

Once the tomatoes, cucumbers, and peppers in the hoop houses started growing, the vines needed to be clipped onto strings hanging

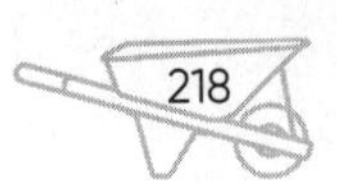

from the rafters to guide the twining, green light-eaters toward the sun. Gro-op members, most juggling other jobs, had trouble keeping up with the work. In search of temporary labor, Walt used his connections at the Mansfield High School to set up a job training program for students where they helped on the farms and earned college credits for botany and agronomy. That worked until summer vacation, when student labor dried up. Walt next went to the local prison, Richland Correctional Institution (RICI). He arranged for a few "honor-roll" prisoners from the minimum-security section to work. The gro-op could not legally pay the prisoners, but helping on the farm qualified as community service work that subtracted time from their sentences. Someone had to get certified to sign out the inmates and bring them back to RICI before the last roll call. Because of the logistics, the arrangement wasn't ideal, but steadily the beds got weeded and the plants were strung. White, purple, and red radishes shot up along rows of raised beds like Indy 500 racers. Swiss chard alternated in its stoplight fashion, green, yellow, and red.

Harvest time came, and the growers delivered the produce, washed and boxed, as contracted, to the gro-op's large walk-in fridge. But in that first year, several buyers reneged on about half their orders. Gro-op farmers returned a week later with more produce, but last week's peppers and tomatoes were still in the fridge. Two more weeks passed, and still no movement. The farmers found it deeply distressing to watch the vegetables they had nurtured fade and wilt in cold storage. They got email messages encouraging them to take the harvest to farmers markets and sell it themselves.

What? That was not the deal. How would they make time on top of farming to become retailers?

In the end, the gro-op farmers could not find anyone who would take the produce even for free. Pounds of vegetables went into compost. It was disheartening.

At the meetings, members had some tough discussions. Some disagreements broke out. A few left the gro-op. Those who did best were the farmers working together on the North End Bowman Street farm. They were able to help each other out, pool resources, and share what worked best.

Of course there were hiccups, they told themselves. Any farmer knows that the hardiest plants and the most long-lasting solutions are those that emerge from indigenous soil, organisms that make connections and grow at their own pace. People do the same in their placemaking. As they bungled along, the NECIC cushioned the co-op from financial failure. A few farmers netted $30,000 a year, but some did not even gross $10,000. In the fourth year, 2023, Amanda, working thirty-hour weeks, made $1,500 for the whole season. Her hourly wage amounted to $2.50 per hour, far below minimum wage.

With such meager profits, the co-op could not have managed without a lot of help. Administrators working at NECIC, Ohio State Mansfield, and Mansfield High School provided manpower, expertise, bookkeeping help, and materials. That is the nature of commercial farming. It does not consistently pay off without outside assistance. Since the Great Depression, federal and state budgets have kept farmers afloat with public subsidies, research, and expertise. The public provides more than $38,000 every minute in farm subsidies to American farmers. It makes sense that farmers in cities also need help. As Walt explained, "Urban farms that succeed are the ones that have the whole city behind them."

The Yellow Food Revolution

ALEX MET US AT THE GATE in the tall cyclone fence surrounding the gro-op's main location on Bowman Street. He looked to be in his twenties, small in stature, with translucent skin and eyes that did not lift from the ground. He asked Walt for a job.

"It didn't work out so well last summer, Alex. How about we try it out this year and see if it goes better before I put you on the payroll?"

Alex nodded in agreement.

Walt took him over to a raised bed and gave him instructions. Alex picked up some tools and set to work as Walt and I drove off for a meeting. Walt was shaking his head as he told me about how last summer he would find Alex behind a hoop house, propped up on a folding chair, chilling instead of working.

When we returned a few hours later, Alex was still on the job, sweating, with no hat or shirt, beet red from sunburn. Rudy, an inmate from the local prison, RICI, said he gave Alex sunblock and had him drink some water.

I put on a hat and gloves and headed over to join the guys from RICI working in a hoop house. Both men wore prison khakis and white T-shirts with "DRC Inmate" in bold letters on the back. I pegged tomatoes with plastic clips as Rudy strung lines and Sam cut wire. We all sweated in the clammy atmosphere of a plastic hoop house on a muggy summer day. The guys were playing R&B on the radio. We talked about music. We talked about our kids. We talked about the kinds of things you talk about when you've got a simple, repetitive task before you.

Rudy, deeply tan, short, and broad-shouldered, looked like a fashionable architect in his thick-rimmed, prison-issue glasses. Rudy said he loves to garden. He didn't know that until he started working at the Bowman Street Farm with his best buddy, Sam. Walt did not need to give Sam and Rudy instructions. They arrived in the morning and set to watering. I noticed how they stopped to pull weeds, caring for the beds as if they were their own. They said they liked how they were treated on the farm. Rudy, who just had been clowning around, looked serious for a moment.

"I can be a different person here."

"What are you like at RICI?"

He dropped his chin and puffed his chest: "Get the f—k away from me."

He laughed. "Here, we can say our please and thank you's."

I had a hard time imagining a tough, snarling Rudy. The farm Rudy was charming, warm, and funny.

Walt's truck pulled in, back from an errand. Walt, as usual, rushed about. In the morning, we had stopped by the gro-op's demonstration kitchen and cold storage. Jesse, the crop planner, had laughed when she heard I was shadowing Walt. "You are going to be up till midnight!"

Walt asked the guys from RICI if they wanted to stay later today.

Department of Richland Corrections inmate watering in a Gro-op hoop house, Mansfield.

Rudy did a little dance. "We'll stay as long as you can keep us. Any time away from RICI is good time."

I joined Alex, who was weeding a raised bed. I followed him down the row, planting collard and pea seedlings. We fell to talking. I asked a lot of questions. I learned that Alex had multiple operations and a serious head injury, which is why he's on disability and why he couldn't join the Navy as he wanted to, following his sister. He had grown up in Mansfield. At twenty-four, he'd had a girlfriend who already had a baby when they hooked up. Alex raised the boy as his own, and then after four years, she got a new boyfriend and kicked him out. Now he

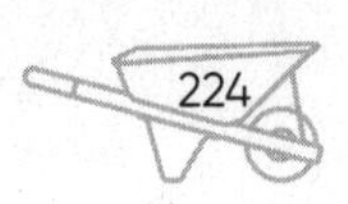

was distressed that he couldn't see the child. He had spent two years in jail because, as he explained it, he was hanging out with some bad people who were dealing drugs, but he was not doing any of that himself. Alex's parents are no longer in the picture, and the grandparents who raised him are dead. Other than his sister and now Walt, who helped Alex find a new place to live when his girlfriend evicted him, Alex had no one to depend on. I got the feeling that Alex came to the Bowman Street Farm as much for human companionship as for the hourly wage to supplement his disability checks.

Later that day when the sun slunk low on the horizon, Walt pulled out two grills and plastic bags filled with ribs, steaks, hotdogs, and, for me, walleye. He grilled corn and asparagus along with the fish, meat, and ribs. Nakeia spread a blue plastic tablecloth on a card table in the large metal shed that served as the farm's headquarters. She then set out fruit, plates, and disposable cutlery.

We sat to eat as the sideways sun turned the shed walls crimson. Walt, who also has a catering business, is an excellent chef. Sam and Rudy joked and ate large quantities of ribs, corn, and steak. Rudy talked about how bad food is in prison, some of it inedible. They dreamed about the first meal they were going to have when they got out. Both men had two and a half years to go.

"I'm not going to eat anything until I have my mother's tamales," Rudy told us.

We cut the watermelon as the sun settled below the horizon. Walt had to get the men back by 9 p.m.

Rudy grinned as he got in the van. "This was the happiest day of my life." Walt smiled just a bit and drove off to take Sam and Rudy back to RICI.

It is hard to express how different the gro-op's local and familial way of doing business is from the large scale of agricultural production in the surrounding central Ohio countryside. Towering over the North End of Mansfield are the largest grain elevators I've ever seen. The big-bellied buildings are topped with voluptuous galvanized steel bins full of corn and soy harvested from Ohio's fertile lowlands. American soy ships abroad to agricultural giants such as China and the Netherlands, where Chinese pigs and Dutch cows gobble up the high-protein feed. Ohio's corn also heads to corn-syrup factories and steer-fattening farms, where animals that evolved to eat grass bulk up quickly on corn. Trying to digest such a diet, cows and steers belch methane in quantities that make up 4.5 percent of total greenhouse gas emissions in the United States. From Mansfield's grain elevators, the vast quantities of corn and soy circulate through feedlots and food factories to return wrapped in plastic in the form of cured meat, chips, cookies, breakfast cereal—a whole grocery cart of processed foods—often of the same yellow color as the Dollar Store logo, the main food store in Mansfield's North End.

How did this resource-heavy food system come about? I puzzled over why gro-op farmers struggled to compete. Why is locally grown, fresh food so expensive? It seemed logical that processed food, produced far away with expensive inputs and packaging, would cost more than potatoes or kale grown nearby, but that is not the case. Cheap, yellow food is a project that has been decades in the making.

In the 1950s, a team of Mexican and American agronomists worked in Mexico to solve the problem of wheat rust. Among them was plant pathologist Norman Borlaug. The scientists developed a new strain of wheat that was not only resistant to rust but also directed large volumes of fertilizer and water to the heads rather than stems of wheat to create varieties that produced up to ten times more grain than conventional strains. The new hybrid seeds triggered a "Green Revolution" across the globe. Wealthy farmers planted new, Mexican wheat in the Americas, Asia, Europe, the Middle East, and Africa. Working through American philanthropists, Borlaug and his colleagues brought to the developing world genetically sophisticated plant varieties that, when paired with state-of-the-art farm machinery, irrigation, chemical

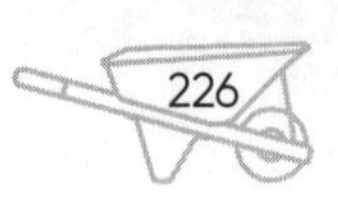

fertilizers, and pesticides, radically increased yields. The Green Revolution made farming less risky and gave food production a clocklike reliability.

In this new system, farmers received public subsidies and private investment to purchase the expensive inputs the hybrid seeds demanded. Large farmers did well. Small farmers had trouble competing, and many went bankrupt. When asked about this tilt of wealth, Borlaug responded dryly, "Our primary concern has to be to produce food. We're not in the business of a land-reform agency." Earl Butz, Nixon's secretary of agriculture, was yet more blunt. He famously told American farmers to "get big or get out."

Since the start of global enclosures of commons two centuries before, people had starved in appalling numbers. Now, the new system worked well to feed urban populations. In the decades that followed, the number of hungry people on earth dropped in half, and the average human lifespan increased by eleven years. In 1970, Borlaug was awarded the Nobel Prize for having saved more lives, allegedly, than any other human in world history. (He didn't do this work alone. Borlaug's Mexican colleagues were overlooked.) Yet in recent years, representatives from the Rockefeller Foundation, which helped fund Borlaug and the Green Revolution, have come to see this story differently. They say the modern food system has "failed" at both nourishing humans and sustaining the planet. Industrial agriculture accounts for 70 percent of freshwater use and at least 25 percent of greenhouse gas emissions, and has contributed to a massive loss of biodiversity. For decades, humans have accepted these environmental losses as the costs of eradicating world hunger.

"I understand the need to produce sustainably," Jeff Lakner, CEO of Lakner Farms, LLC, a major US agribusiness, told me in a phone call in April 2020. "These are things I worry about, but I don't think we can go back 150 years on world trade. How else can we feed nine billion people?"

Lately, however, industrial agriculture has not been doing a great job of feeding people. Growing vulnerable crops dependent on a raft of fertilizers, pesticides, and machinery makes for a fragile system, easily disrupted in recent years by pandemic, climate change, and armed conflict. Since 2017, chronic malnutrition has grown 21 percent. In 2023, 333 million people globally did not know where their next meal was coming from and over 2 billion people, one in four on earth, suffered from malnourishment, micronutrient deficiencies, and obesity. The processed food that Big Ag churns out contributes to a paradox in which many people, especially children, are both overweight and undernourished.

Facing these statistics after decades of backing carbohydrates and animal protein, nutritionists have changed their minds. Vegetables, fruits, and legumes, especially those grown without artificial fertilizers and pesticides, they now say, are better staples of a healthy diet. Yet because state programs award far more subsidies to animal-feed growers than broccoli farmers, only 2 percent of US cropland is planted in fruits and vegetables, which are categorized as "specialty foods."

Mansfield, Ohio, stands at the crossroads of industrial and organic agriculture. In the winter of 1938, the writer and organic farming enthusiast Louis Bromfield bought a farm near Mansfield. When the snow melted, Bromfield was shocked to find his farm was practically worthless. Most of the alluvial soil had been washed off the hills, down the creek, and into the Ohio River. With no topsoil to hold rain, water scoured the hillsides, forming gullies deep enough "to lose your horse inside it." Bromfield set about rebuilding his fields by plowing along the contours of the hills, adding swales to catch rain, rotating and interplanting crops, pasturing livestock for their mineral-rich droppings, and lining the fields with flowering shrubs and fruit trees much like English commoners had two hundred years before. In the decade that followed, as industrial farming took off in Ohio and elsewhere, Bromfield became the country's leading advocate for organic farming. The gro-op picks up on Bromfield's legacy, as members try to take organic farming in a new, risk-sharing direction, toward cooperative enterprise.

Others in Mansfield have come to appreciate the value of the gro-op. The warden of RICI offered the cooperative thirty-four acres of prison land to farm. Deanna and Walt, who had already been teaching business courses at the prison, got into gear partnering with a historically Black university, Central State, to offer the prisoners college courses and job training in urban farming. If they finished the program, inmates graduated with a state-approved certificate in "specialty crop" management. Walt, who barely escaped prison time himself, has made it his business to transform chain-gang farm labor into work that gives incarcerated men a chance at a better life.

Tap roots shoot from the Bowman Street Farm under asphalt, up the hill, through a thicket of brambles to young Alex's lonely apartment. Pale, subterranean filaments radiate from the gro-op under three rows of razor wire at RICI to Sam and Rudy. The roots keep going, spreading to a local food bank, the gro-op's latest customer, and continuing to the corner in front of Amanda's house, which is still a place for drug deals but is also a fern and flower-studded catchment for water running from her tomato-filled hoop house. Roots shoot from her beds and periscope north and south. They reach the neighborhood children who play on Amanda's berms of composted branches, hay, and coffee grounds. "I don't see their parents," she told me, "but I see the kids every day." From above, it's easy to miss the subterranean filaments, but they keep growing. Dozens of community teams from around Ohio flock to Mansfield to learn how to duplicate the gro-op's efforts. My days in Mansfield taught me to wait before I write a place off for its rustbelt woes. As people tend and care for the plants in the city's gardens, other relations between people and across species are also nurtured. Mansfield urban farmers show how tiny garden by tiny garden, they can refashion their world.

THE SEVENTH TURN

Amsterdam

P Is for Plant

THIS WAS SOMETHING I HAD WANTED to do for a long time. I reached down and pulled up a paving stone with a white enamel *P* on it. *P* for parking. I heaved the block onto a pallet in a courtyard in a central neighborhood of The Hague in the Netherlands. Municipal workers would pick up the paving stones, cart them away, and return with organic topsoil. If all went well, no car would ever park on that spot again. I smiled to myself over this small victory. *P*, in my mind, now stood for plant.

Friday mornings, a group met to turn the courtyard parking lot into a garden. The work was going at a snail's pace because, except for me and a forty-something man, DJ from Rwanda, everyone else was over seventy-five. None of the volunteers were interested in gardening, but all of them were happy to see the parking lot ("eye sore") go, to be replaced with fruit trees, flowers, berry bushes, and summer

vegetables. Some of the trees had already arrived in pots with tags giving their Latin names.

This garden was inspired by another one a few blocks away that community activists Marianne Edixhoven and Sacha Winkel had started in 2008 in a vacant lot littered with trash and rubble. They proposed to turn it into a garden and community meeting space. When they signed the lease with the city social housing organization, they were told they could have the space for a year at a time and had to promise to remove any objects left behind after the lease was up.

"It was kind of frightening," Marianne said. "The contract said we would have to pay a fine of 600 euros a day until the site was cleared."

Fearing fines, in the first few years they did not plant perennials or put up any structures in the garden they called the City Oasis Spinoza Hof. But people came to Marianne and Sacha with ideas, desires, and projects. They needed a compost bin and wanted to grow mushrooms. They had a problem with slugs and someone suggested they get ducks. The ducks needed a little pond and a coop. The duck pond quickly filled up with algae, and a gardener engineered an aquaculture filtration system, rinsing the ammonia-rich duck water through gravel beds. The chard and kale planted in the beds grew lavishly, drawing on the nitrogen and potassium the ducks and fish provided. The City Oasis Spinoza Hof had a lot of food growing in the garden, and immigrant gardeners from around the world who were great cooks. A kitchen became another desire. Sacha came up with the idea of creating a kitchen and toolshed from old shipping containers, which they connected with a greenhouse they built with secondhand materials.

The community garden, intended to be temporary, became a commons that has thrived for nearly two decades. On Saturdays, the volunteers cook up a free lunch. On Sundays, they hold a repair café, where people fix neighbors' broken appliances, faulty electronics, and clothing in need of a needle and thread. On Sunday evenings, a choir gathers to give concerts. On weekdays, an over-fifty exercise class meets.

Marianne said she likes to think of the small garden as a talent factory. People come, they have an idea, and they get to work, finding a role to play. The gardeners in The Hague are like the gardeners in Berlin, Tallinn, Mansfield, and Washington, DC. They retain the ability to imagine themselves not as others would have them, but as they want to be. And with that, they unleash astonishing creative powers.

When the neighbors a few blocks away also wanted a garden common, Marianne agreed to help, and so the City Oasis Spinoza Hof garden spread like a blackberry bush that sends out branches that fall to the ground, take root, and grow new stalks. More neighbors appealed to her, and in 2024 Marianne was starting a third garden oasis.

I had come to the Netherlands to live for a year because after the United States, the Netherlands, a country the size of Maryland, is the largest agricultural exporter in the world. Highly urban and densely settled, the Netherlands also has a vibrant urban farming movement. I wanted to understand how people in the Netherlands, where vast stretches of territory are below sea level, are thinking about adaptations in food production to accommodate population density, rising seas, and climate change.

The Netherlands was not always underwater. In the fifteenth century, Holland had vast tracts of boggy peatland. Early modern Dutch farmers excavated peat for fertilizer and fuel. Digging down many meters, they turned land into water. They did this even though people understood that by digging they were sinking their land. But because peat made good money, farmers and entrepreneurs were unwilling to stop. After submerging part of the country, Dutch of all classes built a network of dikes, levees, and windmills to stay dry. Pundits believe the Dutch skill for collaboration and technological innovation comes from centuries of trying to stay afloat.

In the low, wet fields, Dutch farmers could no longer grow grain. They turned to cultivating fodder for livestock, and the Dutch started specializing in cheese. But you can't eat just cheese. By the early seventeenth century the Netherlands became the first country in the world to outsource much of their food production, purchasing wheat and rye from Poland. The so-called Dutch "Golden Age," grew from this global trade. Creaking, wooden tall ships carried enslaved people from West Africa to the West Indies and returned to Amsterdam with tobacco, coffee, cocoa, and sugar. Trading these luxuries, Dutch merchants purchased fine carpets, elegant mansions, and lustrous paintings, the pigments made from expensive minerals mined all over the world.

In the twentieth century, the Netherlands' agricultural power grew. After approximately twenty thousand people died in the Hunger Winter at the end of World War II, the Dutch embraced industrial agriculture. Sicco Mansholt, the postwar secretary of agriculture, led a shift in the following decades from horses and small farms to large-scale, mechanized dairy operations with the latest technologies and chemical fertilizers. Mansholt later became the European commissioner for agriculture and masterminded Europe's common agricultural policy (CAP) designed to lift farmers out of poverty by encouraging them to modernize with the promise of generous subsidies. By the 1980s, CAP had worked so well that European governments spent millions to store a surplus of food—mountains of meat, lakes of wine, and towers of grain and dairy that could not be sold without crashing prices.

The Dutch were all in on the project to mechanize and chemicalize agriculture. They now grow two billion tons of onions and potatoes and a billion tons of tomatoes. Most often this produce is exported, while, strangely, most of the food the Dutch eat is imported. Merchants ship potatoes to Belgium and import potatoes from . . . Belgium. The tubers cross each other on their trade routes. Such are the so-called efficiencies of market capitalism.

But the majority of the crops Dutch farmers grow go to feed not people, but animals. The Netherlands produces fourteen billion kilograms of milk a year, most of it used to make cheese. Every day four thousand calves and twenty thousand pigs are slaughtered. The small country does not generate all these animal products alone. The Netherlands works like a black hole absorbing feedstock from all over the globe. After China, the Dutch buy the most soy in the world, much of it from South America, so that delicious Dutch cheese contributes directly to rainforest depletion in the Amazon.

The imports are essential to keep dairy cows on industrial farms going. A hundred years ago, a farmer fed cows locally grown fodder and would get eight to twelve liters of milk a day seasonally. Today high-performing cows are like Olympic athletes. They decant thirty to sixty liters a day all year round. To keep the milk flowing, farmers purchase high-protein concentrates with ingredients like corn, soy, fish oils, wheat, barley, and sugar pressed into pellets. When farmers feed cows a diet that resembles a fast-food meal, the animals get gassy and suffer from diarrhea and ulcers. They go lame, and their organs shift painfully. Hogs experience a similar punishing physical journey. To gain 270 pounds in just six months, they eat great volumes of soy and corn. A whole shelf of medicines keeps these farm animals going. Farmers treat sickly animals living in crowded conditions with antibiotics to speed growth and fight infections. The antibiotics leach from manure into water sources. They are so ubiquitous that they are causing antibiotic resistance, and helpful bacteria, such as soil-dwelling penicillin, increasingly fail to cure infections in humans.

While much of the milk, cheese, pork, and poultry go abroad, the cow, chicken, and pig manure remain in the Netherlands. The manure contains a lot of nitrogen. Nitrogen, so long desired by farmers, is now a major problem in the Netherlands. Excess nitrogen has degraded soil, water, and air quality. (Amsterdam is the only global city I know that smells like my grandfather's feedlot.) Nitrogen emissions far exceed European Union regulations. The Netherlands leads Europe in biodiversity loss in part due to the great volume of animal poop.

The nitrogen crisis has triggered deep divisions in Dutch society, which has long prided itself on its tolerance for others' views. In 2021, a conservative Dutch government unveiled a plan to spend 25 billion euros to buy farmers' cows in order to reduce nitrogen levels. Farmers exploded in protest. They feared they would go bankrupt without enough cows in production. A newly formed Farmers' Party organized demonstrations and blocked highways with burning trash. Farmers carried banners that read "No farmers, no food" and "There is no nitrogen problem." The ultra-right Forum for Democracy, which supported the farmers, went further, declaring, "There is no climate crisis." At the same time, environmentalists led by Extinction Rebellion, a group organized to compel government action on climate change, blocked highways and hung banners, saying "More Farmers, Less Bullshit." They protested government subsidies for fossil-fuel-intensive industries, including industrial farming.

In 2023, two right-wing parties backed by farmers and industrial agricultural lobbies won parliamentary elections. They sought to withdraw the Netherlands from climate change agreements and several European Union regulations on migration.

In this controversy, the farmers had a point. They were being penalized for longstanding agricultural policy. For decades, agronomists told farmers to maximize production to an industrial scale. States rewarded farmers who complied. Eighty-two percent of the European Union's CAP agricultural subsidies favor resource-intensive, animal-based foods at the expense of vegetables, fruits, and nuts. Billions of euros in public support make meat and dairy cheap, everyday goods, even though producing and consuming these foods in excess is extremely costly to both public health and environments. Insurers, banks, supermarkets, and the food-processing industry also reward large-scale animal husbandry. Asking farmers to suddenly cut back or transition without requiring all the other players in the production chain to adjust was, indeed, unfair.

Extinction Rebellion activists also had a point. Nitrogen pollution and climate change are real. Livestock production contributes heavily to both problems. "No farmers, no food" is accurate, but the animal husbandry industry is pretty piggish. It requires 70 to 80 percent of the world's agricultural land to produce only 18 percent of calories and 25 percent of all protein consumed by humans. Despite a new plant-based food movement, the global demand for meat is on the rise. That is no surprise. Agricultural policies, technologies, and subsidies have trained modern consumers to eat animal products, which have become cheaper and more readily available since the animal husbandry revolution of the 1980s. In the Netherlands, despite pledges to reduce nitrogen and carbon, the government and corporations are largely continuing with business as usual. Like Dutch farmers centuries ago who were unwilling to stop digging peat, the flawed food system contributes to sinking humans into a heating caldron. Something has to give.

What if one essential consumer good—fresh food (fruits, vegetables, nuts)—came not from over a thousand miles away, as it does on average today, but from a fifty-mile radius? Switching to local, sustainable food production would presumably punch a hole in greenhouse gases and could lower the cost of food, since shipping, storing, and packaging make up the bulk of the price at the cash register. Might there be other benefits, too? Would the simple act of pulling food back home have an impact on citizens, health, and politics?

With these questions in mind, I went to work at an urban farm called Pluk, on the western edge of Amsterdam. Pluk fed about 150 families who subscribed and showed up weekly to pick their share of the harvest. Three paid farmers supervised several dozen volunteers and interns on two and a half acres.

Because land in Amsterdam is expensive to rent or buy, Pluk relied on garden beds scattered across an organic orchard. These were wastes, edges that the orchard owner could not use. The gift of affordable farmland was terrific, but the layout made for extremely inefficient farming. We spent a lot of time and energy hauling wheelbarrows and tools from one distant bed to another. Pluk farmers earned, at best, a minimum wage. Interns and volunteers were paid in vegetables. Even so, no one shirked the work. As in Mansfield, we did simple, repetitive tasks that took little thought. We would squat on the ground under a hot sun or a chilling rain, pulling weeds and planting seedlings down long rows. I felt the difference in my body between the scales of gardening and farming. The farm labor was quietly exhausting, causing a dull aching press on my knees and lower back. To ease the tedium, we talked as we worked. We took long tea and lunch breaks where we shared food and feelings, celebrated birthdays, and took note of disappointments. I came to find my place as a foreigner in Amsterdam thanks to the farm.

Until the middle of the twentieth century, a ring of truck farms like Pluk had circled Amsterdam. The city's waste streams, including human waste, nurtured the growth of the beloved potatoes, kale, and cabbage of the Dutch diet. By 2023, only two farms were left, tiny

Pluk farmers hauling compost, Amsterdam.

Pluk and its partner, a six-acre farm down the road. But there, at the Boeterblum farm, trouble was brewing.

Through some shady deals for which two officials went to jail, the city had acquired the farm's parcel of land and some 103 acres surrounding it called the Lutkemeerpolder. The territory had once been a lake (*meer* in Dutch), formed from excessive peat digging. In 1664, L. H. Rutgers van Rozenburg invested money from his sugar plantations in Suriname to pay workers to dig a network of dykes and windmills to enclose what had been a public lake and turn it into a private farm. In a strange irony, the people living near the Lutkemeerpolder today were mostly migrants from Suriname. The labor of their enslaved ancestors had created the wealth that turned lake back into land and enclosed it.

In the 2010s, city leaders sought to make another transition. They wanted to convert the farmland into an industrial park, planting, instead of carrots, six-acre gunmetal gray warehouses on the last stretch of Amsterdam's organic farmland.

In September 2020, construction crews started building the industrial park. A small group of protestors gathered and stood in front of the bulldozers. Police arrived. In a video clip from that time, a tall policeman walked up to an older woman and shoved her to the ground. Police arrested others. With the path cleared, the bulldozers pushed on.

In the spring of 2021, activists started a diggers' rebellion. With shovels and seeds, they began to construct garden beds on the Lutkemeerpolder. They spaded with a particular design in mind. Each garden bed was shaped in a letter of the alphabet spelling their rallying cry. From planes taking off from Schiphol, passengers could read "Save Lutkemeer!" written in flowers and vegetables on the earth below them. In August, just when the produce from the alpha-beds was ready to harvest, the police ordered the gardeners to leave the site.

Many protestors refused, and several people set up tents and camped. After a week, Amsterdam police repeated the actions of authorities in Berlin a century before. They moved in to pull down the camp and garden. Months of work was wiped away in just a few hours.

Under pressure, city leaders scrambled to justify the industrial park. There were already empty warehouses in the city looking for business. City leaders announced they had a client willing to lease the logistics center planned for the Lutkemeerpolder, but they would not disclose who the partner was.

Activists went through public documents attached to the deal. The name of the silent partner was blacked out, except for in one spot where they found the name of AH, Albert Heijn, a large grocery store chain. They announced to the press that the popular franchise had entered an agreement with the city-owned development concern to pave over the Lutkemeerpolder. Albert Heijn did not want that fact disclosed because the company marketed itself as environmentally friendly.

The story ran in the press for a day before the busy public clicked on to the next headline. Activism in the infinite world of social media requires a great deal of creativity. Protestors realized that most Amsterdammers did not take notice of events in the working-class margin of the city. So, they decided to focus their next action in the city core.

On September 18, 2021, employees went to open the doors at Albert Heijn stores. In neighborhoods across Amsterdam, managers could not get their keys into the locks. As crowds of shoppers gathered, news went around that locks on the doors of eighteen Albert Heijn stores had been glued shut.

Later in the day a group calling itself "AH Must Go" released a statement explaining the protest action as an attempt to stop the grocer from building "a large concrete distribution center in the precious Lutkemeerpolder."

Two weeks later, the grocer canceled the lease for a warehouse on the Lutkemeerpolder.

But even with no partner, city leaders kept going. They insisted they needed to build a big-box distribution center on organic farmland to earn revenue for social welfare.

I spoke to Natascha Hulst, a sustainability consultant involved in the movement. The economy itself, Hulst observed, was the problem. The main issue blocking the transition to sustainable food production was access to affordable land. In Amsterdam, land speculation had driven prices sky-high. Why should markets rule and not people, Hulst wondered. She thought about the commons, which had for centuries given Europeans access to land and resources.

"I finally realized we don't have to be crushed between the market and the state."

Hulst and others founded an organization and drafted a project to turn the Lutkemeerpolder into a "food park," an agroecological landscape for community gardens, commercial urban farming, and a makerspace where residents could organize themselves. Instead of a logistics center to facilitate the transportation of food across long distances, they aimed to create healthy jobs and grant residents a role in producing more affordable food with less food waste. They called their new organization "Voedselpark [Food Park] Amsterdam."

The Food Park initiative shifted the conversation. The cause was no longer about blocking development, but offered a vision for community wealth building on the Lutkemeerpolder and perhaps for the city as a whole. Hulst calculated that they needed 4.3 million euros to buy the 103-acre site to turn it into a land trust. They raised money through crowdsourcing. Small donations rolled in, amounting to four hundred thousand euros, enough for a down payment.

While the meetings and protests continued, I kept showing up for my work shifts at Pluk. The farm had no tractor or plow. No shovels went into the ground. We used only a broad fork (picture a large pitchfork) to aerate the soil and hand tools to slip out weeds. One aspect of nourishing the soil is leaving it in peace. For the communities of soil microorganisms, plowing is warfare. Digging exposes microbes to air and they die. Plowed soils that have no fungi and microbes fall apart. They dry out and roll downhill or take off in the wind. Farmers then add fertilizers on depleted fields to keep plants growing. It becomes an addiction; exhausted soils require more and more treatments in the form of fertilizer to keep them going.

Instead, we hauled compost to beds around the farm after harvesting each crop, laying down an inch or two. It took me a while to understand what we were doing. Compost is not soil. Nor is it fertilizer. The decayed leaf mulch we used does not have much in the way of mineral composition to feed plants, but it serves as a shelter and a feeding trough for microbial life in the soil. In healthy tilth, microbes work with plants. Plants use carbon dioxide, water, and sunlight to produce sugar. But plants, like humans, cannot live on sugar alone.

Pluk farmers Gen and Edu in a patch of kale, Amsterdam.

They need nutrients too. Fungi and microbes serve as tiny miners in the earth. They take atoms of minerals in the ground and turn them into chemical forms that plants can use. In exchange, plants give them sugar. As we balanced heavy wheelbarrows of mulch down narrow paths, we were not feeding plants directly this or that mineral. We were spreading a banquet for our tiniest, microscopic workers on the farm.

Pluk operates on a small budget. Its biggest expense, after salaries, was purchasing composted leaves from the city, which came at a hefty price. We often talked about the problem of not having enough compost to layer on the beds while so many valuable organic materials—food scraps, spoiled produce—that could be turned into compost went into landfills. In the Netherlands, less than 2 percent of discarded organic resources are captured for reuse. In 2021, 2.8 million tons of food were thrown away. If you piled up all that rotting produce, it could fill ten skyscrapers. Valuable nutrients are all around us, if we could just recognize them.

Fides Lapidaire handed me a postcard with the headline "Let's Give a Shit." Under it was a photo of Fides, pants down, squatting in a Dutch pasture, windmill in the background. The young artist has a defiant look on her face.

"I love poop!" Fides told me over lunch.

Fides had converted a horse trailer so that it has two openings, one leading to a food truck, the other to a toilet. She sets it up at festivals. Visitors who used the composting toilet could walk around the trailer to the food truck counter and buy discounted sandwiches made with vegetables that Fides grew from earth fortified with the sandwich-eaters' bodily donations. The sandwiches, grilled eggplant and zucchini, looked delicious. People eagerly lined up for the exchange.

Fides thanked people roundly for their fecal gifts to make the point that human feces is a collection of rich mineral resources sucked up from fields around the world, barged into Amsterdam, eaten, and ejected from human bodies. Instead of throwing this wealth away, she wants to mine it. Fides and her partner Yanna Hoek hold "Give a Shit" festivals each spring. They made a documentary film, "Holy Shit," and regularly speak to groups of farmers, agricultural suppliers, policymakers, and politicians. As crazy as it sounds, two long-legged young women peddling the values of poop are starting to have an impact. After their last "Give a Shit" festival, several parliamentarians raised the question of rethinking the country's waste streams.

"It will take a long time." Fides laughs.

A lot of artists make art, speak truth to power, and move on. Fides was not moving on. She created a foundation and raised some money.

In 2024, she and her partner were working on a new project, Feel-Gut, a supplement designed to feed your gut in a healthy way so you can become the best poop producer possible. They were in the process of setting up a new alliance with a portable toilet company, some farmers, and a food-processing company to produce both compost and food. To do so, they will have to challenge Dutch laws that ban the use of human waste in food production. Fides is undaunted.

"Here were all these huge environmental problems, and I suddenly realized I could do something to help correct it. I could take a dump! We can all produce good fertilizers. We do it anyway."

Of all sources of animal waste, human feces best feeds soils that grow food for humans. That is not surprising. Nutrients that nourish humans churn through bodies and come out the other end. In the Netherlands, precious agricultural minerals flushed down the toilet include seventy-seven million pounds of phosphate and seventy-four million pounds of potassium a year. Agronomists worry that mining concerns will soon run out of phosphate in the coming decades. Fides is also concerned.

"There are plates full of nutrients in your poop that are now being washed away and burned without looking back."

Fides hopes that by installing composting or vacuum toilets (like in airplanes) in new buildings, cities could conserve drinking water and help farmers overcome an addiction to artificial fertilizers made from petroleum products. Such commercial fertilizers supply plants with the basics they need to grow—nitrogen, phosphorous, and potassium (NPK), plus some salt. If you taste fertilizer, it is very salty, like drinking seawater. Pumped up with salt, plants grow tall, and that makes farmers happy, but the produce contains a lot of water and is low in nutrition. Likely you have tasted large, but bland, watery tomatoes, peppers, strawberries, and celery. These fruits and vegetables do not inspire much culinary interest. Humans evolved to distinguish instinctively plants that are deep in color, fragrant, and rich in taste because they contain nutrients that best feed human bodies.

If people eat nourishing foods and compost them back into the ground, the same nutrients are now in the earth, feeding soil organisms that in turn cycle them back to plants that produce the roots, seeds, and fruits that sustain human health. Fides's point is simple. Close the loop and suddenly every human is a producer of soil and plant food, and our worlds become so much richer. If you have read this far, you probably already know that soils with deep humus that are rich in fungal and microbial life store both water and carbon with astonishing efficiency. All you have to do, Fides is saying, is give a shit.

Globalism Under Glass

IF YOU CAN SOLVE THE PROBLEM of nourishing soils, a person can get on to growing food. In the Dutch city of Almere there is an allotment garden the likes of which I had never seen before. It spreads inside a massive greenhouse, about the size of a city block. Within the glass walls, 1,500 gardeners rent plots for serious money. Most community gardens charge about $50 a year. Onze, which means "ours" in Dutch, charges $25 a month for a room-sized plot. When I visited the allotment under glass with Jan Eelco Jansma, an agronomist from Wageningen University, the majority of renters were from the Caribbean, working-class migrants who earned on the low end of the pay scale. One gardener, Denis, had pooled together a dozen plots and cultivated a fifth of an acre. He had towering banana trees, pole beans, and tomato plants that climbed on strings twenty feet into the air. On the ground, he grew eggplant, bok choy, and Surinamese greens, bitter leaf and callaloo. Denis had a living room set up, too, with a couch, teakettle, coffee table, and a chalkboard listing produce and prices. Two women, one Dutch, one Surinamese, had stopped to shop and joke around with him. I asked Denis if he made a living with this garden.

"No, not at all. I sell these things to pay the rent for all these allotments. That is $350 a month."

"Why do you do it?"

"I love it."

"Do you spend much time at it?"

"No, it doesn't take much time," Denis said, clearly itching to return to his plants, "unless these ladies come here and bother me."

Denis, laughing, nodded toward the two friends who had made themselves comfortable on his couch.

Ron van Zwet runs the greenhouse community garden. Initially, he grew roses in the industrial greenhouse, but van Zwet had trouble competing with cheap imported roses and high-volume growers in automated greenhouses. He wondered what else he could do with the space.

When he came up with the idea of dividing the greenhouse into allotments and renting them out to gardeners, none of his fellow farmers thought it was a good idea. Almere is a mainly working-class city with a large contingent of migrants. Who would pay that much to grow a few vegetables that you can buy cheaply in chain stores? Other farmers predicted the greenhouse would quickly be overrun with pests and blight brought in by people who did not know what they were doing. Even so, van Zwet thought it was worth a try. Waiting lists for outdoor allotments were growing, while the number of allotments were shrinking by about 10 percent in two decades of feverish urban development around Dutch cities.

For each plot, van Zwet offered water, electricity, fresh compost, minimal heating (above freezing), and organic pest control, mostly in the form of wasps and other insects that eat aphids and moths. For five years he worked two jobs while the greenhouse allotment got going. He printed leaflets in Dutch but had few takers. Finally, his children advised him to advertise in the Surinamese newspaper. He did, and word got around quickly. Rentals took off.

Almere is the newest Dutch city. Just over fifty years ago the territory where the city stands was a matrix of triangular waves sloshing back and forth across the Zuiderzee, a large shallow bay of the North Sea. In 1972, the Dutch carried out their most ambitious and last land reclamation project by damming the entire bay and digging a network of ditches to dry the sea bottom formerly submerged beneath several meters of water. The next step involved figuring out what to do with the land appropriated from fish, aquatic plants, and seabirds. Most of the new acres went to farming in an era of agricultural overproduction.

On the reclaimed land, planners also plotted out two new cities, Lelystad and Almere. In Almere, architects were inspired by Ebenezer Howard's vision of a garden city, and drew up plans for parks, boulevards, and plenty of green space (but few gardens) among housing towers, residential subdivisions, and suburban streets. But the new housing sat empty. Dutch people who had a choice did not choose to move to Almere. The Dutch liked their crowded, charming old cities with tilting houses on cobbled streets. Almere was too new, too artificial, too sprawling. People on budgets and migrants ended up moving there.

These were the people who showed up at van Zwet's greenhouse willing to pay high prices for small plots. Surinamese gardeners were interested in growing vegetables from the Caribbean—bitter melons, hot peppers, tropical spinach, bananas, okra, and solanum nigrum. Vegetables imported from Suriname were expensive. Thanks to Onze's

Caribbean-like climate, Surinamese migrants could grow fresh, tropical vegetables year-round. The warm, moist air in the greenhouse the day that I visited must have felt a bit like home for Dutch-Surinamese living in a country that trends gray and cool.

As word got around, more and more people signed up. Van Zwet created a waiting list for plots. He extended the greenhouse by a third. And he built a second branch in a nearby city. The new allotments were also quickly rented.

Van Zwet kept the gardens supplied. In a little store, he sold seedlings, seeds, and vegetables. If you did not know about gardening, van Zwet would plant and tend your crops for an extra cost. Jan Eelco and I met people who had no garden but strolled along in the greenhouse as if it were a park. Some would stop to buy produce from gardeners.

The greenhouse was breathtaking. Most greenhouses have concrete floors with plants growing on tables. In Onze, gardeners planted directly into black earth. They used every inch of space to grow a world of plants. In early spring, Dutch cabbages, potatoes, and pear trees sprouted next to Surinamese mangos, vining tropical greens, and peppers. Here was a botanical globalism I had never experienced before.

Jan Eelco observed, "I think they have no major pest problem because of the biodiversity. It's the large greenhouses growing just one or two crops that really suffer."

We stopped and talked to a Dutch gardener. His small plot was stuffed with a dozen varieties of peppers, a half dozen kinds of tomatoes, plus climbing beans, zucchini, cucumber, and melon vines crawling up the netted walls of his tiny domain. The big man stepped gingerly around his plants.

"What do you do with all this produce?"

"We eat it. I preserve the tomatoes and make hot sauces out of the peppers. My neighbors are from Suriname and Indonesia. They taught me to like spicy food. I give them all they want, and my son comes and takes away vegetables every week to his family."

I understood the pride in the man's voice, that feeling of being able to feed people he cared about.

"Does it take much of your time?"

"I come every other day and spend about two hours."

"Do you ever buy produce in the grocery store?"

The man laughed, and looked around him. "No, no need for that."

"People are always telling me that gardening is hard work. What do you think?"

"No, it's not work. I do it for fun. And I like knowing where my food comes from."

Growing food can be hard work. It takes effort to set up a garden, and it is no fun at all when your plants are sickly and refuse to grow. Some days when I was volunteering at Pluk, the weeding never seemed to end. But growing in small spaces, like the gardeners do at Onze, is much easier. Plants clustered tightly together don't leave much room for weeds. There's little trudging or hauling. According to a time-budget survey of gardeners in Soviet Russia, people spent from thirty minutes to six hours a week in their dacha gardens, though surveyors had a hard time calculating labor hours because gardeners showed up not only to work but to visit with their plants and neighbors.

In addition to the work/not-work issue, another question nagged me. Are small urban gardens like those in Onze a more sustainable way to produce food than conventional agriculture, which can grow in great volumes?

I found that was not an easy question to answer. A 2022 study looked at the difference in yields between conventional and urban agriculture of all types—peri-urban farms like Pluk, backyard gardens, and one-hundred-million-dollar vertical urban farms run with robots. This meta-analysis of over a thousand papers found that in most categories of crops, urban agriculture was similar to or had greater yields than conventional agriculture. Some crops like tomatoes and fruits were two to four times more productive.

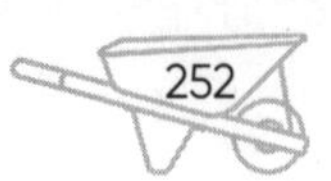

Yields, however, do not say much about inputs and sustainability. Another study found that the urban farms that outperformed conventional agriculture were indoor hydroponic farms, which burn through a lot of carbon in heating, lighting, and equipment, smoking out up to six times more CO_2 than conventional agriculture. But farms and individual gardens that had existed for a long time outperformed conventional agriculture in both yield and carbon emissions. Urban agriculture in some countries, like Poland, where there is a long history of allotment gardening, scored far better than conventional agriculture on both emissions and yields. At Pluk, farmers and customers arrived on bicycles. We did the work manually. Other than deliveries of some seedlings in plastic containers and a truckload of compost every few years, Pluk was powered without fossil fuels.

Urban planners estimate that cities can at best meet only about a third of food demand. Noncommercial outdoor farms and gardens with the smallest carbon footprint, scholars postulate, are not those that would have a major impact on feeding cities, at least not on today's diet. Gardeners in the past, in Berlin, DC, and Soviet cities, managed to cultivate most of what they ate because they had a largely plant-based diet supplemented with some eggs, poultry, and pork. But people don't eat that way anymore. They mostly consume foods that demand a lot of land, water, and resources—meat, grain, sugar, and dairy. These resource-intensive foods are difficult to produce in cities.

The good news is that researchers find that when people start gardening, they change their diets to eat more fruits, nuts, and vegetables. Could revamped food-growing systems inspire a large-scale transformation in how people eat? What if, in a future world, farmers grew food in a way that not only reduced carbon emissions but sequestered carbon and stored water? What if growing food could also restore biodiversity?

Wouter van Eck makes his office in the village community center amid pensioners playing bridge. He doesn't own a car and he walks to his farm, which is difficult to distinguish as a farm. Wouter cultivates a food forest on six acres in the village of Groesbeek in the eastern part of the Netherlands. A food forest is a productive food system designed by humans and modeled after a natural forest. It is like a parking garage that instead of layering cars, stacks edible plants floor by floor. Drivers start at the garage roof, a canopy of tall nut-bearing trees, and descend to a lower level of fruit trees. They go down another story to shrubs, and yet lower to herbaceous ground cover before slipping into a basement rhizosphere of root crops and fungi.

In an orderly checkerboard landscape of green pastures, roads, and canals, Wouter's square looks like a hairy, crazy mistake. Weeds line the road. A narrow muddy trail takes us across a bridge where a small stream flows. Willows and seagrass crowd the brook. A few beavers, once extinct in Holland, have taken up residence in the canal edging Wouter's food forest. We stepped over saplings the animals had felled. I raised an eyebrow. What farmer would want beaver in a farming concern that is dependent on trees? Wouter just grinned in delight, pointing out the trail the beavers had made to their lodge.

Wouter had beavers, but other things were missing. I saw no barn, tractor, or irrigation system, no tool shed, not even one tool. It is hard to even see crops on this farm. Wouter insisted they were there. The farm has over four hundred species of plants, two hundred of them edible. He ran through a list of fruit trees: medlar, pawpaw, fig, apricot, pear, sumac, quince, apple (many varieties), Mongolian lemon, plum, almond, and Chinese mulberry. As Wouter listed his fruiting perennials, I was reminded of the commoners' hedge in an early modern English village, which peasants designed as a living pantry. Wouter pointed to a poplar that had a berry-producing chocolate vine running up it, about forty feet tall. The only leaves on the tree during one visit in November were those of this vine with tiny red-brown flowers.

The chocolate vines, like the edible ground cover (hosta, ostrich fern, wild garlic [ramsons]), grow quickly in spring before the tree leaves bud. Then they chill in the shade for the rest of the growing season. To generate the shade that the ground cover needs, Wouter planted fast-growing, short-lived trees. The poplars and willows also provided a canopy for saplings that like a deep shade. Wouter stopped at a Korean pine, an evergreen no taller than a person. It and the towering poplar beside it were planted at the same time. The poplar will die in forty to sixty years (if Wouter doesn't cut it sooner) and the Korean pine will last for 450 years, producing pine nuts most every year.

There were, as everywhere in Holland, stinging nettles, which Wouter picks, heats to remove the sting, and eats as greens, but, he pointed out, insects and caterpillars love them too. They would rather eat nutritious nettles than the leaves of Wouter's apple trees.

I marveled at the complex matrix of botanical, insect, and animal life of this tiny forest. Wouter is known as the founder of a growing food forest movement in the Netherlands. I asked how he came up with the idea.

He explained that in the late 1980s, he had an internship in Kenya researching Dutch development projects there. In one village, Dutch consultants were introducing hybrid corn on a slope. Plowing caused erosion, which forced the African farmers to go into debt buying fertilizer and pesticides. During a dry period, the crops didn't grow well. Heavy rains came, and more soil slid downhill. Wouter watched the farmers fall deeper into debt and despondency.

"I was a bit younger and maybe a bit more rude, and so I started quarreling with the people implementing this project because it was not helpful. Both the farm community and nature were suffering. One of them got irritated and said, 'Well, what do you want—that over there?' And he pointed to where a lush forest was growing, but also people were living there."

Wouter crossed the valley and saw that growing amid the small households were mango, avocado, and papaya trees. Bananas grew in the understory, and below them, shade-loving coffee, tea plants, and a whole lot else, edible perennials from across Asia and South America. The forest was noisy with birds and insects and had the rich scent of moist earth. These households did not struggle with their crops. They were not going into debt to buy expensive seeds, fertilizer, and pesticides imported from abroad. Why, Wouter wondered, were European aid agencies trying to export their broken agricultural system to people who had built a healthy, multicultural society of crops?

Wouter returned from Kenya and got on with his life. Decades passed in which Wouter shifted from working as an environmental activist to winning a seat as a city councilman with the Green Left Party. He became well-known, infamous even, for trying to ban SUVs from the city of Nijmegen. (In response, angry car drivers took out life insurance policies in his name.) He successfully stopped a plan to bulldoze a seventeenth-century convent to build a parking garage. As time passed, Wouter thought more and more about the food-filled Kenyan forest.

In 2009, Wouter decided to become a farmer. He bought a conventional agricultural field, where for decades farmers had plowed in the fall and left bare earth. In spring they would break the soil again, sow seeds, and add fertilizers and pesticides to grow fodder for animals. In late summer, they harvested and left the earth exposed again. Wouter changed course radically. He did not mow or plow. The first fall he planted willows and poplars as a windbreak. During the long winter evenings that followed, he designed his food forest. He drew up a color-coded map. Blue for water, yellow for sun-loving plants, and green for those that prefer shade. He researched plants from all over the world, embracing cultivars of all kinds, native and migrants. Finally, spring arrived, and with the help of friends and volunteers, he started planting.

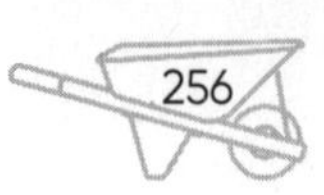

The problems started soon after. In the first year, a plague of green caterpillars descended, eating every leaf from his beautiful, blossoming plum trees. Shifting from environmentalist to farmer tested Wouter's politics.

"I started hating caterpillars."

As a politician, he had campaigned for years against chemical pesticides, but as a farmer Wouter considered different organic pesticides that would exterminate the caterpillars.

"But at the same moment, a certain skill in me developed which I call intuitive laziness. Many problems in life stop, if you are patient enough, by doing nothing."

That very spring, insect-eating birds arrived from Southern Europe and Africa. They were rare species that had been disappearing from agricultural landscapes. They come to the Netherlands to nest, lay eggs, and use the long summer days to forage food for their chicks. Wouter figured that if he killed the caterpillars, the red-listed birds would have nothing to eat. He pitied the birds, and so let the caterpillars stay. The plum trees struggled and produced little, but the following year more rare birds flew in and with them butterflies, beetles, and bird watchers. The birds devoured the little green caterpillars, and the plum trees thrived.

Caterpillars are one thing, but beavers are another. Farmers across Europe and North America kill beavers and blow up their dams because they flood fields, clog waterways, and kill trees.

Wouter's beavers had swum upstream from Germany, passing miles of agricultural fields with nothing but fodder crops. Wouter's willows were like a roadside diner. The beavers stopped and set up shop.

As they chewed their way along the bank of the canal, Wouter worried. He consulted the web to learn about beavers. He discovered that beavers prefer to eat willow and that the bulky, short-legged animals do not travel overland far from the stream. Wouter had been thinking it was time to thin the wildly growing willow along the canal. As an intuitively lazy farmer, the decision was easy. He decided to leave that job to the beavers.

So, the beavers stayed. They built a dam on the canal, a small pond swelling out behind it. At night the beavers swam along another canal and dammed up a stream running through the middle of the lot, creating another small pond. Felled trees lay across it; dozens of willow saplings sprouted on its edge. Beavers work a bit like Wouter. They carry willow branches here and there, messily leaving them on banks where the branches take root and sprout into saplings. Wouter plants trees to harvest later. Beavers appear to be doing the same.

As we walked along the bank, Wouter looked at the beaver pond with satisfaction.

"The prediction is for a dry April, and if this coming summer is like the last three years, we won't have much rain."

Beaver ponds slow the flow of water, giving it time to drip into underground reservoirs. Come summer, the roots of Wouter's trees will draw from that water table. Wouter showed aerial photos of his food forest during the record-dry summer of 2022. The image of his green forest next to neighbors' jaundiced fields of hay and alfalfa starkly spotlighted the difference in resilience. While the fodder crops wilted and died in the sun, the trees' deep roots drew water from the depths and carried it up through the length of their vascular network like giant straws sucking water toward leaves that in turn sprayed a fine mist in the forest. These solar-powered fogging machines leave the air wetter and cooler. Neighboring shrubs and plants take in that moisture, which helps them survive droughts too.

We stood above the once straight, human-dug ditch watching the stream burble over the beaver dam. The water on the downstream side was noticeably clearer after passing through the dam's reedy filter. I was struck by the fact that I had never seen clear, fresh water in the

Netherlands before. I tried to imagine the volume of phosphate and nitrogen that could be recovered from behind those dams if beavers had the habitat and the permission to return to the Low Countries.

A bird let out a sharp, short call. Wouter pointed into a thicket.

"That's a pheasant warning other birds that a predator is around."

Suddenly the chorus of birdsong went silent as the small birds waited till the shadow of the buzzard flying overhead departed. Then they started up again, spotting one another in chirps and warbles under a brilliant sun on a cool spring day.

"We have thirty species of birds here." Wouter smiled his crooked smile.

I understood the birds' attraction to this sheltering forest stocked with berries, fruits, seeds, worms, insects, and amphibians. The food forest is a supermarket for birds. Wouter, who when I dropped by served up warm drinks and tasty snacks, treats the birds with the same generosity. Unlike most orchardists, he does not prune his trees. They are healthier, he insists, growing tall with all their branches intact.

"So we harvest the fruits and nuts we can reach with a ladder and we leave the rest for the birds."

Wouter's aim in seeking to reproduce the biodiversity and self-fertilizing fecundity of a wild forest is to grow food without fossil-fuel inputs. And that he does. We stopped under a chestnut tree, three stories tall, which had just produced its first good crop of the sweet, protein-rich nut. Wouter estimates that a hectare (2.5 acres) of food forest in a temperate climate can feed about ten people, far more than industrial agriculture does in an area the same size. Scientists generally agree with this assessment.

Wouter plucked a leaf of hogweed, considered, as the name suggests, a pernicious weed, and gave it to me to try. It tasted like kale infused with herbs.

Fine, but who will eat hogweed for dinner?

To answer that question, I left Wouter to head to Nijmegen with a few friends to have dinner at De Nieuwe Winkel, a Michelin-starred restaurant that in 2022 was voted the best vegan restaurant in the world. De Nieuwe Winkel serves only food grown within fifty kilometers. The restaurant staff visit Wouter's food forest every Monday to harvest. We sat down to a meal of mustard seed, white Dutch truffle risotto (as we were in Holland, the risotto was made of sunflower seeds instead of rice), pickled radishes and skoby, a pâté of beets and walnuts, tartlets with chopped wild garlic, hogweed, mushrooms, and seaweed. For dessert, rhubarb, stripped and raw, with almond ice cream and freshly pressed almond oil (which tasted like something alive, like eating a delicious, liquified nut).

As the evening wound down, the restaurant's chef, Emile van der Staak, showed us around.

"When we started, we had empty tables for a long time."

No one was interested in eating food grown within a few miles in a small country with a narrow range of biodiversity. But gradually word got around that Dutch food could be more than kale, potatoes, and sausage. Van der Staak and his colleagues shifted the menu toward "plant-based" foods (*vegan* being a politically charged word, they avoided it). Some diners, realizing they had just had a meal with no dairy or meat, refused to pay.

"We had some interesting conversations over that," van der Staak said, but he didn't mind disgruntled customers.

"I don't do this for a Michelin star."

Rather, he sets his sights on the goal of designing a modern food palette that would enable people to eat delicious food from their locality without causing harm to themselves, their ecologies, or people and landscapes in other parts of the world.

"We do our part," van der Staak said with an animated gleam in his eye. "We could scale this up. There is nothing to stop us from eating bread made from chestnut flour."

Imagine that. The wheat fields that first pushed peasants from their commons, which caused so much urban misery, the same fields that today add great puffs of carbon to the atmosphere and consume underground mountains of fossil fuels and rivers of water, could be replaced by towering groves of chestnut trees, a new bread of life that enriches human bodies, and much more. Chestnut forests with shade-loving edible plants growing in their understory would spread across Holland. As the forests matured, their considerable volume of leaves and branches would decompose to build soil. Low-lying, flood-prone Holland would elevate each year millimeter by millimeter, reaching one day the level it had been before Dutch farmers dug out the peat, sinking their land below the sea. Heavy rains, rather than rushing through dikes to flood towns and contaminate lakes and seas, would pool behind beaver dams and filter slowly to underground reservoirs. The trees each day would breathe in carbon and exhale oxygen. People would be drawn to those forests for rest, quiet, and the scents—relaxing volatile organic compounds—that plants emanate. Birds, insects, and animals would come to the woods for their own purposes. Farmers like Wouter would have good reason to welcome them.

CONCLUSION

The Diggers

IN 1649 IN ENGLAND, dispossessed commoners started a rebellion. Defying enclosure, they climbed a hill, pulled out spades, and started digging up gardens. The Diggers left behind their declarations. They claimed commons and wasteland as a common treasury, something they could all work and enjoy in search of collective sustenance. The Diggers proceeded, shoveling up one colony at a time, spreading peacefully to other regions. Political theorist Susan Marks writes that the Diggers laid no claim to the land of the rich, although they saw the wealthy and their greed as a major social problem. They declared: "The rich could remain in their enclosures, saying This is mine," while the poor lived "upon their Commons, saying This is ours." Digging with simple spades, Amsterdam activists in the twenty-first century joined their seventeenth-century brothers and sisters, spading up the "trade of darkness" whereby the rich accumulate wealth "under the cover of law."

For the past two hundred years, peasants and small farmers who lost their rights to food, fuel, and shelter moved to cities where they recreated commons and regenerated urban wastes. They dug up hard-packed earth, recycled discarded nutrients, and turned vacant land into lush, fertile landscapes. Working-class gardeners built with their hands the idea that cities could be more than stone and brick, more than markets and factories. They demonstrated that by turning their part of the city into a garden, they could get by, even flourish, without exploiting others somewhere else. As they worked, urban farmers made use of the river of organic materials that flow daily into metropolitan areas. Turning garbage into soil, they devised the most productive agriculture in recorded human history. And there is more to it than food. Working-class gardeners showed the way to the idea of garden cities. Their mutual aid societies inspired the first glimmers of the social welfare state. As this history has shown, when people exert their rights, take ownership of their resources, and find a way to belong to their landscapes, they shore up civil society, devise resilient economies, and shape more inclusive citizenship.

But what comes next, after the gardens and communities supported by them flourish? In Berlin, Paris, Tallinn, Amsterdam, and Washington, DC, urban farmers with their organization and labor added value to real estate. Their tenure on the land was provisional and short term, and others were eager to swoop in, capitalizing on their labor and ingenuity. From nineteenth-century Berlin to twenty-first-century Amsterdam, gardeners had a simple request. They asked for land to be returned to commons—to be placed outside real estate markets so that people could cultivate with confidence that their labor and creativity would not be stolen. Repeatedly, the competitive demands of commerce overruled them. For this reason, small-plot self-provisioning thrived best in this history in the Soviet Union and East Germany, where gardeners had access to public commons and where the state encouraged tiny gardens with botanical creations such as Michurin's hardy trees and shrubs and regulations (even if regulations in general were excessive). State support and public land gave gardeners staying power.

Now the need to decrease CO_2 emissions opens new spaces for tiny gardens to again flourish. Cities in the Netherlands are gradually shedding their automobiles. Central Amsterdam allows access only to delivery trucks, cabs, and permitted residents. Planners arrange one-way streets so that it takes a very long time to drive across the city, while public transit and bike traffic flow freely. Utrecht is turning an industrial park into the first car-free neighborhood in the country. The Hague, a Dutch city with unusually wide streets and many parking lots, is gradually trimming pavement away.

That's the marvel of cities and their suburbs. They can be extremely dynamic congregations for human communities to live and work because metropolitan areas are always in a state of flux as people reuse and reinvent infrastructures. Bicycles and small electric people-movers (such as scooters, electric wheelchairs, and e-bikes) are very efficient forms of human transportation, and they don't take up much space. The phasing out of gas-burning automobiles will liberate public spaces once devoted to parking and moving cars. These commons could be sold to developers, as city leaders sought to do with the Lutkemeerpolder in Amsterdam, or they could instead be dedicated more profitably to plants and gardens. Like interwar Estonian cities, major streets could become boulevards of nut and fruit trees with berries in the understory.

Citizens could ask for a change in municipal zoning so that, for example, a new condo would have, not a parking space, but so many square feet of garden assigned to it. Suburbanites would be free to grow food in their front yards, and streetside parking could become curbside allotments, while parking lots could turn into cooperative gardens. Given the long waitlists for a plot in a community garden, it is safe to predict that people would flock to these fruit and vege-

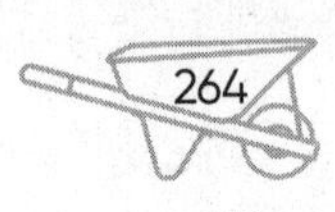

table oases. Neighbors could elect a garden coordinator to educate and organize gardeners: gifted leaders like Kate Tully, the cooperative garden organizer in Washington, DC; Walt Bonham, the manager of Mansfield's Gro-op; and Marianne Edixhoven in The Hague. Parks and weedy edges could be transformed into food forests. Warehouses and big-box stores might be capped with heat-trapping, water-recycling greenhouses for all-season community allotments like Onze in Almere. As landscape architects Thomas Rainer and Claudia West imagine it, "The next renaissance of human culture will be the reconstruction of our natural world in our cities. Plants will be the center of it all."

You might scoff at the suggestion that tiny gardens could help solve the planetary-scale problems of both climate change and a looming "polyfood crisis." But keep in mind that in post-Soviet countries in the crisis decade of the 1990s, citizens in tiny gardens grew from 60 to 90 percent of the food they ate on 1.5 percent of the arable land with few fossil-fuel inputs. Climate change is one of many problems that feels too big to tackle for any one entity, be it the US government or UN agencies. But it would not take an international convention or an act of Congress to make room for tiny urban gardens. Just a local zoning change would do it.

In a city dappled with gardens, people would be out on the streets walking, biking, rolling, or scootering, and they would be tending their plots, harvesting fruit in season, enjoying the cleaner, cooler air that the plants provide. People with their hands in the soil would exchange produce with one another, swapping human and soil microbes, creating kin and communities. In cities with edible landscapes, people would likely be healthier. Food scientists report that fresh food grown in organic soils can have 25 percent more nutrients than produce from industrial fields. Studies show that people who live near green spaces, especially fruiting shrubs and flowers, have a healthier gut microbiome than people who do not. Horticultural therapists find that children and adults who take part in gardening therapy recover faster than those who don't. Other studies show that neighborhood greenness is associated with lower rates of obesity and decreased risk of stroke and heart disease. (Despite the association of green space with good health, however, rural people, living in areas clouded with pesticides, farm runoff, and nitrogen contamination, do not enjoy this botanical benefit. After a century of being healthier than city people, since the 1980s, rural dwellers now die younger, a trend epidemiologists call a "rural mortality penalty.")

But there is more than just individual health at stake. People in cities can work with other organisms to build up healthy ground that provides food and habitat for birds and animals who increasingly seek refuge in cities from polluted agricultural spaces. Meanwhile, people who know where their food comes from tend to waste less and share more.

It would not always be easy. Farmers at Pluk in Amsterdam struggled in the summers of 2018, 2019, and 2022 with drought, and in the summers of 2023 and 2024 with excessive rain. Urban farmers may need to rely more on perennial plants with more resistance, as in Wouter van Eck's food forest. They will also benefit from organized recycling of gray water, food scraps, leaves, clippings, and the rich nutrients from composting toilets to build fertile, hardier, human-engineered soils. Circling organic wealth through cities, people soon could become like early human ancestors in mega settlements who, it appears, did not need to conquer other people's territory, extract resources abroad, or maintain hierarchical or racial states that forced some to work and kept others out. That was true because gardening differs greatly from farming. While industrial farmers declare war on the environment, gardeners work with microbial, insect, and animal allies to turn the hard work of farming into the cornucopia and community of gardening. This more harmonious story could be not just in the deep past, but in the future, one that draws on the uniquely rich resources of the modern megalopolis.

Over the past three centuries, small gardens worked like tiny coral worms that create a shelter and join with other worms to construct a reef that draws a world of fish, plants, and color to its architectural marvels. Such, too, is the creativity of the dispossessed. It falls to them to seek ways to build, protect, restore, and imagine more just and equitable relationships. The stories in this book induce wonder and offer new ways of seeing, and new possibilities. They serve up hope for restoration and peace, one tiny garden at a time.

Acknowledgments

I do not have artificial intelligence to thank for this book, but I do have many people who generated collective knowledge. First, I want to thank a friend who can no longer read this book. Warren Cohen has been my mentor, cheerleader, and supporter for the last quarter of a century. Thank you, Warren. And thank you Marianne Szegedy-Maszak for caring for him as if you really were his daughter.

There are many people who allowed this book to come alive and for whom meeting them was so joyful. I will start with the gardeners. Wesley Cornell taught me nearly every complicated thing I know about plants. He patiently answered my questions, even when he had already answered them before. Ryoko Yamamoto showed me by example how to feed the earth so that the plants are happy. I thank Nicole Virgil for the long talks, delicious lunch, and her unflinching conviction and integrity. Georgette Norman in Memphis and Denise Morrison in Tulsa generously answered questions when a stranger called out of the blue (and then called again). In Mansfield, I want to thank the farmers

Walt Bonham, Nakeia Nickelson, Vincent Owen, Amanda Standfield, Mona Kneuss, Tim Hicks, Marqua Meyers, and Jess Hudson. They are awesome. The farmers at Pluk—Gen, Edu, and Illy—made me feel welcome from the start. I watched as they did that for everyone, drawing people in, surrounding them with their care. Ann Doherty got them started, leading with a quiet wisdom and infectious optimism. In The Hague, Marianne Edixhoven let me tag along. Wim Goris invited me, a perfect stranger, to The Hague just so I could meet Marianne and see Erasmus's former kitchen garden. In DC, Sasha Bamford Brown and Marjoleine Kars helped me haul buckets and harvest kale. Sasha interpreted from Spanish, built maps for this project, and hunted down data in the DC Recorder of Deeds and the US census. His work made a huge difference. Dave Bamford, Kama Garrison, and Leila Corcoran taught me about permaculture and a whole lot else. Leila also turned her hoe on this text, culling it fruitfully. Michelle Feige with her usual generosity let me design her garden and experiment with it while she hosted me on her farm. Kate Tully gave me an unforgettable tour. Patricia Browne opened her school and gardens to me. Josh Singer and Kate Love helped me navigate the DC garden administrative world. The gardeners in Estonia are an impressive group managing short summers and rocky ground with nonchalance. My thanks to Tiuu Saarist, Sirje Viita, Riste Laasberg, Tiina Ann Kirss, Peeter Veegen, Mart Pungo, Henn Sokk, and Enu and Anu Printsman. They all welcomed me, showed me around, and gave me apples. Svetlana Trofimova and Olena Palko told me with touching honesty their stories of hardship gardening in the 1990s. I admire them both very much. Dorothee Wierling, who has cultivated my mind since she was one of my advisors in the 1990s, was always ready to serve up wine, fresh produce, and lots of laughter in her Charlottenburg Kleingärten in Berlin.

Moving from growers to guides, I want to thank Sarah Tarlow, who, when I asked for a tour of the open-field English landscape, answered my email in minutes and eventually invited me, whom her daughter called her "internet friend," to stay over at her place in Grantham. Linda Kaljundi in Estonia met me with beautifully detailed lists of people she thought I might like to get to know for this project in Estonia. Linda took me to gardens around Tallinn and farther. In her quiet way, she led me on a journey, each open door more fascinating than the last. Aljona Suržikova gave me a tour of the airport gardens and introduced me to her mother-in-law, Svetlana Trofimova, whom I liked immediately. Epp Lankots drove me to Paide, interpreted from Estonian, and taught me about dacha architecture. Anneli Banner, a garden historian, gave me a fascinating tour of collective farm life. Deniss Polubojarov showed me around Narva's dacha complexes and even gave me documents to take home. Kadri Tüür and Ulrike Plathe helped me explore Estonian culture. In DC, Bernadette James befriended us in her café and showed us around Deanwood. We learned how she cultivates children as well as flowers. Steve Coleman gave us a tour of Deanwood's community gardens. In Holland, Jan Eelco Jansma gave up a valuable day to show me Almere. Wouter van Eck generously gave me his time when I came by. The inspired and passionate Anna Kooi showed me the way in all things related to food. In Berlin, the garden history team of the Weissensee allotment welcomed us with tea and cookies, books, photos, and wonderful stories. Iris Poels led me on tour after tour of the many facets of the Dutch alternative food movement. Iris is a genius at knowing which stories to tell when.

I relied on many people for help with sources. Rosa Oefinger worked with me steadfastly and brilliantly. Caio Rossant helped with some seriously legal translations. Gilberto Mazzoli sent me very useful USDA materials from his collection. Maraya Taylor provided me with a hard-to-find dissertation. Amanda Seligman gave me advice and help with sources. Susan Oosthuizen led me on an amazing tutorial on the ecology of the fens. My thanks to Samir Megheli and Jennifer Morris at the Anacostia Community Museum and Shelly Buring at the George Washington University Special Collections for sending materials. At the Netherlands Institute for Advanced Study (NIAS), Elly Theelen and Erwin Nolet found the books I needed so efficiently. A number of MIT students helped with this project, doing amazing things with databases. They include Brooke Dodrill and Kim Cheng, plus all the students in our digital humanities project. The graduate students in my seminars on plants, wild and cultivated, helped me think my way through the scholarship of the plant humanities. Eve Allen and Brandon Scott shared both their impressive botanical knowledge and their concoctions.

Many people read drafts of this project. My thanks to Nancy Ries and her colleagues at Colgate University. At NIAS, I am grateful to Isobel Schneider, Jan Wilhelm, Zara Kars, and Bernike Pasveer, who made our visit so lovely, and to the fellows, especially Lotte Hoek, who generously offered their friendship. My thanks to Johanna Conterio, Susanne Bauer, and Ursula Munster for inviting me to Oslo to a workshop on soil. Olha Martynyk, Benjamin Schenk, and their colleagues at Basel University helped greatly with the Soviet chapters. Derek Musgrove, my sweet friend, read and commented on the DC chapters. No one else could have done that with such knowledge. Tom Fleischman and Sarah Tarlow did the same for the chapters on Berlin and England. I am grateful to my colleagues at MIT who are supportive in so many ways. Megan Black is always willing to listen to me sort out ideas. Kenda Mutongi took time from her busy life to come to DC to help me out. She cheerfully walked up and down streets with me, exploring and finding ways to have us laugh. Her presence made all the difference.

This book has been masterfully curated by my agent Sarah Lazin and editor Alane Mason. I benefit greatly from their years of experience and foresight. Thanks to YJ Wang at Norton for shepherding *Tiny Gardens* through the press and to Illy Kerr, an artist and farmer, who gave so much thought and care to the illustrations for this book.

Finally, thanks to my big, extended family. They are always asking to read what comes next and happy to share gardening advice and photos of garden beds. For me, they are the first mutual aid society. My sister Julie Hofmeister applied her laser-sharp designer eye to this project. Marjoleine Kars, my partner of several decades, first hears out every thought and then reads every word, often more than once. I am grateful for her patience, love, and loyalty.

Notes

INTRODUCTION

2 **a coming "polyfood crisis":** On polyfood crisis, see David Wallace-Wells, "Food as You Know It Is About to Change," *New York Times*, July 28, 2024.

3 **stitched together urban ecosystems:** For information on urban ecosystems, see Ben Wilson, *Urban Jungle: The History and Future of Nature in the City* (Doubleday, 2023), 198; Peter S. Alagona, *The Accidental Ecosystem: People and Wildlife in American Cities* (University of California Press, 2022); and Erica Spotswood, Erin Beller, Robin Grossinger, J. Letitia Grenier, Nicole Heller, and Myla Aronson, "The Biological Deserts Fallacy: Cities in Their Landscapes Contribute More than We Think to Regional Biodiversity," *BioScience* 71, no. 2 (February, 2021): 148–60.

4 **They came together to rebuild commons:** For information on commons, see E. P. Thompson, *Customs in Common* (Penguin, 1993), 126.

4 **people who cared for and defended urban green space:** Dawn Bichler, *Animating Central Park: A Multispecies History* (University of Washington Press, 2024).

1. LOCKED OUT

12 **contact with soil microbes:** For the benefits of soil microbes, see Anirudra Parajuli et al., "Yard Vegetation Is Associated with Gut Microbiota Composition," *Science of the Total Environment* 713 (April 2020): 136.

13 **City dwellers piped in fresh water:** On city dwellers' connection with the environment, see William Cronon, *Nature's Metropolis: Chicago and the Great West* (W. W. Norton, 1992).

13 **poorly paid job in a factory:** For information on working conditions in factories, see National Assistance Board, *Reports of Special Assistant Poor Law Commissioners on the Employment of Women and Children in Agriculture* (London: W. Clowes and Sons, 1843), 65–88.

13 **1830, Kent, England:** For information on social unrest in Kent, England, see John E. Archer, *Social Unrest and Popular Protest in England, 1780–1840* (Cambridge University Press, 2000), 22.

14 **communally managed property:** For more on commons, see Garrett Hardin, "The Tragedy of the Commons," *Science* 162, no. 3859 (December 1968): 1243–48. Hardin repeated arguments of the conservative landowner, William Forster Lloyd, *Two Lectures on the Checks to Population: Delivered Before the University of Oxford, in Michaelmas Term 1832* (Oxford: S. Collingwood, 1833), 30–1; and Christopher Dyer, Erik Thoen, and Tom Williamson, "Conclusion: The Rationale of Open Fields," in *Peasants and Their Fields: The Rationale of Open-Field Agriculture, c. 700–1800* (Brepols, 2018).

15 **Ostrom on commons:** Elinor Ostrom, *Governing the Commons: The Evolution of Institutions for Collective Action.* The Political Economy of Institutions and Decisions (Cambridge University Press, 1990).

15 **what an open agricultural system looked like:** For information on sowing commons, see Gilbert Slater, *The English Peasantry and the Enclosure of Common Fields* (Constable, 1907), 48, 53.

16 **"nursery for weeds":** Arthur Young, *An Inquiry into the Propriety of Applying Wastes to the Better Maintenance and Support of the Poor* (London: J. Rackham, 1801), 20.

16 **English landowner and novice farmer:** For information on Jethro Tull, see Laura B. Sayre, "The Pre-History of Soil Science: Jethro Tull, the Invention of the Seed Drill, and the Foundations of Modern Agriculture," *Physics and Chemistry of the Earth, Parts A/B/C* 35, no. 15–18 (2010): 851.

17 **Commoners wanted no part:** On the struggle against industrialization, see Bruno Caprettini and Hans-Joachim Voth, "Rage against the Machines: Labor-Saving Technology and Unrest in Industrializing England," *American Economic Review: Insights* 2, no. 3 (September 2020): 305–20.

17 **Saving seeds for the new season:** Gary Paul Nabhan, *Where Our*

Food Comes From: Retracing Nikolay Vavilov's Quest to End Famine (Island Press/ Shearwater Books, 2009): 52.

18 **went into the pot for the daily soup:** For information on edible plants used for soup, see Margaret Willes, *The Gardens of the British Working Class* (Yale University Press, 2014), 9.

18 **vast tracts of land called "wastes":** For information on "wastes," see Sarah Tarlow, *The Archaeology of Improvement in Britain, 1750–1850* (Cambridge University Press, 2007), 44.

18 **turned to wastes to enrich the family diet:** On diets in industrial Britain, see Chris Otter, *Diet for a Large Planet: Industrial Britain, Food Systems, and World Ecology* (University of Chicago Press, 2020).

19 **"overstocked and ill-kempt lands":** as quoted in Jane Humphries, "Enclosures, Common Rights, and Women: The Proletarianization of Families in the Late Eighteenth and Early Nineteenth Centuries," *Journal of Economic History* 50, no. 1 (March 1990): 25. On commoners drawing on the commons "for the whole of their miserable existence," see John Sinclair, ed. *General Report on Enclosures Drawn up by Order of the Board of Agriculture* (London: McMillan, 1808): 17.

19 **after he paid for horses, fertilizer, and labor:** On wages from a kitchen garden, see Humphries, "Enclosures, Common Rights, and Women," 17–42.

19 **pre-enclosure hedges made up lateral food forests:** For information on pre-enclosure hedges, see Philip J. Wilson, "Botanical Diversity in the Hedges and Field Margins of Lowland Britain," in John W. Dover, ed., *The Ecology of Hedgerows and Field Margins* (Routledge, 2019), 2–22.

19 **On collectively managed, renewable forestry,** see William Bryant Logan, *Sprout Lands: Tending the Endless Gift of Trees* (W. W. Norton, 2019); and Andrew S. Mathews, Trees Are Shape Shifters: How Cultivation, Climate Change, and Disaster Create Landscapes (Yale University Press, 2022).

20 **"less inclined to work":** As quoted in J. M. Neeson, *Commoners: Common Right, Enclosure and Social Change in England, 1700–1820* (Cambridge University Press, 1993), 28–29.

20 **"at least with as little as possible":** As quoted in J. L. Hammond and Barbara Hammond, *The Village Labourer, 1760–1832* (Longmans, Green, and Co., 1911), 10.

2. THE DUNG COMMUNITY

21 **Peasants broadcast offerings:** For information on the veneration of the dung heap, see George Ewart Evans, *The Pattern Under the Plough* (Faber and Faber, 1971), 125.

22 **"folded" or penned them by night:** For information on folding sheep, see Slater, *English Peasantry*, 114, 24.

22 **"folding quality":** Neeson, *Commoners*, 69.

22 **a fly has already planted her eggs:** For information on insects' role in digesting dung, see Peter Skidmore, *Insects of the British Cow-Dung Community* (AIDGAP, 1991), 3–35, 141–47.

23 **to get the soil chemistry right:** For information on fertilizer use and its effect on soils, see Humphry Davy, "On the Analysis of Soils," in *Communications to the Board of Agriculture, on Subjects Relative to the Husbandry and Internal Improvement of the Country* (W. Bulmer & Company, 1805), 303–18; and Greta Marchesi, "Justus von Liebig Makes the World: Soil Properties and Social Change in the Nineteenth Century," *Environmental Humanities* 12, no. 1 (May 2020): 205–27.

23 **initially increased harvest yields:** See tables in M. E. Turner, J. V. Beckett, and B. Afton, "The Wheat Question," in *Farm Production in England, 1700–1914* (Oxford University Press, 2001).

23 **farmers purchased more fertilizers:** For information on increased use of fertilizer, see Turner et al., "The Wheat Question," 20.

24 **These were "Luddites":** For information on Luddites, see Richard W. Hoyle, *Custom, Improvement and the Landscape in Early Modern Britain* (Routledge, 2011), 1; and Neeson, *Commoners*, 30.

24 **cost savings, and greater profits:** For information on the adoption of commercial agriculture, see John Sinclair, *The Code of Agriculture: Including Observations on Gardens, Orchards, Woods, and Plantations* (London: Sherwood, Neely, and Jones, 1821).

24 **"If you offer them work":** as quoted in Hammond and Hammond, *Village Labourer*, 9.

24 **"bring children and eat cake":** as quoted in J. D. Chambers, "Enclosure and Labour Supply in the Industrial Revolution," *Economic History Review* 5, no. 3 (1953): 319–43.

25 **"only hunger can spur":** As quoted in K. D. M. Snell, *Annals of the Labouring Poor: Social Change and Agrarian England, 1660–1990* (Cambridge University Press, 1985), 123.

25 **"labourers will work every day":** Hammond and Hammond, *Village Labourer*, 9.

25 **The discipline of wage labor:** On views toward wage labor, see Jeremy Burchardt, *The Allotment Movement in England, 1793–1873* (Boydell Press, 2014), 79–80.

26 **so much red tape:** Author telephone interview with Kate Lee, June 23, 2020.

28 **"no one could claim a right . . . to deprive it of life":** Susan

Marks, *A False Tree of Liberty: Human Rights in Radical Thought* (Oxford University Press, 2019), 173. For more on Spence, see 176–80.

29 **Prime Minister William Pitt asked:** On William Pitt's views toward agriculture, see William Pitt, *General View of the Agriculture of the County of Northampton* (London: Richard Phillips, 1809), 62.

29 **the social impact of the new round of enclosures:** Robert C. Allen, *Enclosure and the Yeoman: The Agricultural Development of the South Midlands* (Oxford University Press, 1992), chapter 9.

29 **tags pinned to their chests at hiring fairs:** On English hiring fairs, see Michael Roberts, "'Waiting Upon Chance': English Hiring Fairs and their Meanings from the 14th to the 20th Century," *Journal of Historical Sociology*, vol. 1 (1988).

30 **"refuse gainful employment":** Snell, *Annals of the Labouring Poor*, 124.

30 **"are kept 'off the books' ":** Jason W. Moore, "The Capitalocene, Part I: On the Nature and Origins of Our Ecological Crisis," *Journal of Peasant Studies* 44, no. 3 (May 2017): 594–630. On the creation of new words in the sixteenth century, *private property* and *pauper*, see Steven Stoll, *Ramp Hollow: The Ordeal of Appalachia* (Hill and Wang, 2017), 55, 57.

30 **"speak up in their defense":** Young, *An Inquiry into the Propriety of Applying Wastes*, 51.

30 **he suddenly changed course:** On Arthur Young's change in views toward the poor, see Snell, *Annals of the Labouring Poor*, 111; and "Agricultural Report," *Sunday Times*, January 8, 1832, Issue 481.

30 **"no use is made":** Snell, *Annals of the Labouring Poor*, 1.

31 **three-quarters of an acre:** Snell, *Annals of the Labouring Poor*, 12–14.

31 **"excess populations":** "Debates of Parliament," *Sunday Times*, March 14, 1830, Issue 386.

31 **"home colonization":** Jeremy Burchardt, *Allotment Movement in England*: 76. For more information on home colonization, see Neeson, *Commoners*, 76–8. For allotment lessor testimony, see "Debates of Parliament," *Sunday Times*, March 14, 1830, Issue 386.

31 **"allotment dukes":** For information on "allotment dukes," see Burchardt, *Allotment Movement in England*, 101–2.

31 **hated the idea of allotments:** For information on landowners' opposition to allotments, see Jeremy Burchardt, "Rural Social Relations, 1830–50: Opposition to Allotments for Labourers," *Agricultural History Review* 45, no. 2 (1987), 165 75.

31 **"a dangerous scheme":** Robert Hyde Greg, "The Allotment System," *Economist*, November 16, 1844, 1426.

31 **"the less they work for us":** David Crouch and Colin Ward, *The Allotment: Its Landscape and Culture* (Faber and Faber, 1988), 49.

3. ALLOTTED SURVIVAL

35 **Parliament passed three acts:** On allotments for self-provisioning in England, see Micheline Nilsen, *The Working Man's Green Space: Allotment Gardens in England, France, and Germany, 1870–1919* (University of Virginia Press, 2014), 22; Burchardt, *Allotment Movement in England*, 232–33; John E. Archer, "The Nineteenth-Century Allotment: Half an Acre and a Row," *Economic History Review* 50, no. 1 (February 1997): 23; Arthur Wilfred Ashy, *Allotments and Small Holdings in Oxfordshire* (National Institute of Agricultural Engineering, 1917), 3.

36 **On elites not returning the commons:** Montague Gore, *Allotment of Land: A Letter to Landed Proprietors, on the Advantages of Giving the Poor Allotments of Land* (James Ridgway, 1831): 6; and "Debates of Parliament," February 28, 1830, *Sunday Times*, Issue 384.

36 **"grow their own poor rates":** Crouch and Ward, *Allotment*, 100.

36 **"insufficient for a full subsistence":** George Field, "Hints for the Cultivation of Peat-Bogs in Ireland with a View to the Increase of the Population, Security and Happiness in that Part of the United Kingdom in a Letter to T. Rev. Malthus," in *A Brief Outline of the Universal System* (1816), 92; and Fredrik Albritton Jonsson, *Enlightenment's Frontier: The Scottish Highlands and the Origins of Environmentalism* (Yale University Press, 2013), 12.

36 **landlords ruled allotments:** For more on allotment rights, see Archer, "Nineteenth-Century Allotment," 21–36; Humphries, "Enclosures, Common Rights, and Women."

37 **"allotments than going to jail":** Archer, "Nineteenth-Century Allotment."

37 **"symbol of dispossession":** Crouch and Ward, *Allotment*, 61.

37 **soil was gray and dry:** On the effects of the agricultural revolution on soil, see tables in Turner et al., "The Wheat Question." On deterioration of fields, see Vaclav Smil, *Enriching the Earth: Fritz Haber, Carl Bosch, and the Transformation of World Food Production* (MIT Press, 2001), 39.

38 **For more on the Little Grand Canyon in Georgia:** Paul S. Sutter, "What Gullies Mean: Georgia's 'Little Grand Canyon' and Southern Environmental History," *Journal of Southern History* 76, no. 3 (2010): 579.

38 **the problem of degraded soils:** Pëtr Kropotkin, *Fields, Factories and Workshops* (Thomas Nelson & Sons, 1912): 61. For more about nineteenth-century thinkers on the degradation of soils, see Hannah Holleman, *Dust Bowls of Empire: Imperialism, Environmental Politics and the Injustice of "Green" Capitalism* (Yale University Press, 2018), 103–4. On Marx's understanding

of the "metabolic rift," see John Bellamy Foster, "Marx's Theory of Metabolic Rift: Classical Foundations for Environmental Sociology," *American Journal of Sociology* 105, no. 2 (September 1999): 366–405.

38 **"the poor man's crop":** Stephen Demainbray, *The Poor Man's Best Friend, or, Land to Cultivate for His Own Benefit* (London: J. Ridgway, 1831), 16.

38 **"rarely a successful or permanent cultivator":** Ashy, *Allotments and Small Holdings*, 54. On higher yields with spade farming, see Jonsson, *Enlightenment's Frontier*, 21.

39 **spade farming led to higher yields:** On gardeners' yields, see Burchardt, *Allotment Movement in England*, 4–6; Jonsson, *Enlightenment's Frontier*, 21.

39 **to reach a half million plots:** On the growth of allotments in the nineteenth and twentieth centuries, see Crouch and Ward, *Allotment*, 61; Malcolm Chase, "'Wholesome Object Lessons': The Chartist Land Plan in Retrospect," *English Historical Review* 118, no. 475 (February 2003): 59–85; Burchardt, *Allotment Movement in England*, 213; Denis M. Moran, *The Allotment Movement in Britain* (Peter Lang, 1990), 35–37.

40 **a connection to others:** David Graeber and David Wengrow, *The Dawn of Everything: A New History of Humanity* (Farrar, Straus and Giroux, 2021), conclusion.

41 **2,000 pounds of food:** Author interview with Rosie Williams, November 14, 2020, Washington, DC.

42 **"always food left over":** Author interview with Kate Tully, Columbia Heights Green, May 3, 2021, Washington, DC.

4. THE GREEN GHETTO

45 **warned travelers to steer clear:** On nineteenth-century working-class neighborhoods in Berlin, see Max Ring, "Ein Besuch in Barackia," *Die Gartenlaube* 28 (1872): 458–461; Theodor Fontane, *On Tangled Paths: An Everyday Berlin Story* (Angel Books, 2010), 147; "Baracken vor den Toren," *Illustrirte Zeitung*, June 15, 1872; and Johannes Böttner, "Die Laubenkolonien," in *Der Praktische Ratgeber im Obst und Gartenbau* 19, no. 43 (1904): 393–96.

47 **built palatial villas:** For information on Berlin's industrial boom and new food technologies, see Alexandra Richie, *Faust's Metropolis: A History of Berlin* (Carroll & Graf, 1998), 164.

48 **population multiplied beyond all historical precedent:** For information on changes in land tenure and population growth, see Marion W. Gray, "Urban Sewage and Green Meadows: Berlin's Expansion to the South 1870–1920," *Central European History* 47, no. 2 (2014): 275–306;

David Blackbourn, *The Conquest of Nature: Water, Landscape, and the Making of Modern Germany* (W. W. Norton, 2006), chapter 3.

48 **spoke of the "housing crisis":** "Kurze Analyse der Wohnungsfrage im Rahmen einer preußisch-österreichischen Konferenz im November. 1872 zur sozialen Frage," https://ausstellungen.deutsche-digitale-bibliothek.de/preussen-reichsgruendung-1871/items/show/39, accessed March 14, 2023. For information on Berlin's housing crisis, see Belinda Davis, "'Everyday' Protest and the Culture of Conflict in Berlin, 1830–1980," in Andreas W. Daum and Christof Mauch, eds., *Berlin–Washington, 1800–2000: Capital Cities, Cultural Representation, and National Identities* (Cambridge University Press, 2005), 263–85; On 15,000 Berliners without housing, Alex Vasudevan, *Metropolitan Preoccupations: The Spatial Politics of Squatting in Berlin* (Wiley/Blackwell, 2016), 27.

49 **thousands of people took up residency:** For information on the "Barrack Republic," see Catharina Paetzelt, *Die Ersten 100 Jahre: Die Verbandsgeschichte Des Deutschen Kleingartenwesens* (Bundesverband Deutscher Gartenfreunde, 2021), 26–27; Nilsen, *Working Man's Green Space*, 74; *Neuer Social-Democrat*, July 28, 1871, in Johan Friedrich Geist und Klaus Kürvers, *Das Berliner Mietshaus. 2, 1862–1945: Eine Dokumentarische Geschichte von "Meyer's Hof" in der Ackerstraße 132–133, der Entstehung der Berliner Mietshausquartiere und der Reichshauptstadt Zwischen Gründung und Untergang* (Pressten-Verlag München, 1984), 104; Kurt Wernicke, "Barackia," *Berlinische Monatsschrift* 8 (1997); Annemarie Lange, *Berlin zur Zeit Bebels und Bismarcks: Zwischen Reichsgründung und Jahrhundertwende* (Dietz, 1972), 133–34.

49 **"with that special German thoroughness":** Ring, "Ein Besuch in Barackia."

49 **"provided for its basic necessities":** "Die Baracken vor den Toren," *Illustrierte Zeitung*, in Geist and Kürvers, *Das Berliner Mietshaus*, 111.

49 **structures were to be "arbors,":** On social conditions and regulation in Berlin shantytowns, see Manfred Kassel under Günter Landgraf, *Zur Geschichte des Organisierten Kleingartenwesens in Berlin-Treptow* (Vorstand des Bizerksverbandes der Gartenfreunde Berlin-Treptow e. V., 2005), 6; Eva Brücker, "Und Ich Bin Heil Da 'Rausgekommen': Gewalt und Sexualität in Einer Berliner Arbeiternachbarschaft Zwischen 1916/17 under 1958," *Physische Gewalt: Studien Zur Geschichte Der Neuzeit* (1996): 337–65; Kristin Poling, *Germany's Urban Frontiers: Nature and History on the Edge of the Nineteenth-Century City* (University of Pittsburgh Press, 2020), 120–27.

50 **"I'm supposed to pay 90 Thalers for a cellar hole":** Ring, "Ein Besuch in Baracki a."

50 **Peat turf absorbs moisture:** On building materials, see Rita Gudermann, "Mitbesitz an Gottes Erde- Die Ökologischen Folgen der Gemeinheitsteilungen," *Jahrbuch für Wirtschaftsgeschichte / Economic History Yearbook* 41, no. 2 (January 2000): 85–110.

50 **These "Schreber gardens":** For information on the Schreber gardens, see Heinrich Hinz, *Der Schrebergarten: Praktische Ratschläge zur Einrichtung und Bewirtschaftung von Schreber-, Klein-, und Hausgärten* (Frankfurt an der Oder: Trowitsch & Sohn, 1915).

51 **"torments of the cosmopolitan city":** Ring, "Ein Besuch in Barackia."

52 **journalist called Barackia an "idyll":** Geist and Kürvers, *Das Berliner Mietshaus*, 111. On a neighborly willingness to help, see Johannes Böttner, "Die Laubenkolonien," in *Der Praktische Ratgeber im Obst und Gartenbau* 19, no. 43 (1904): 393–96.

52 **in place for ten years:** Hanna Hilbrandt, *Housing in the Margins: Negotiating Urban Formalities in Berlin's Allotment Gardens* (Wiley, 2021), 34.

55 **an experiment in self-government:** On the Paris Commune, see Kristin Ross, *Communal Luxury: The Political Imaginary of the Paris Commune* (Verso, 2015), 73.

55 **Thousands attended meetings in the summer:** On political organizing in Berlin's shantytowns in 1872, see excerpt from "Auszug aus dem Bericht über die grosse Volksversammlung vom 23.7," *Neuer Social-Democrat*, July 28, 1871, in Geist and Kürvers, *Das Berliner Mietshaus*, 103–104.

56 **the newly established "building police":** On outlawing dwellings in Barackia, see Wernicke, "Barackia."

56 **"landlords so arrogant":** "Revolte in Berlin," *Neuer Social-Demokrat*, July 31, 1872. On real estate speculation, see Otto Glagau, "Häuserschacher und Baustellenwucher," in Geist and Kürvers, *Das Berliner Mietshaus*, 120. On uprisings in July in Barackia, see "Bericht des Polizei- Lieutenant Rath," and "Bericht des Stellvertretenden Polizei-Präsidenten," July 27, 1872, in Geist and Kürvers, *Das Berliner Mietshaus*, 114, 117–118; Davis, "'Everyday' Protest and the Culture of Conflict," 270; and *Neue Preußische Zeitung*, July 30, 1872.

58 **"a formal guerrilla war":** "Bericht des stellvertretenden Polizei-Präsidenten," July 27, 1872, in Geist and Kürvers, *Das Berliner Mietshaus*, 117–118.

58 **The rest were "outsiders":** "Pöbelexcesse," *Spenersche Zeitung*, July 31, 1872.

59 **a stay of eviction:** *Spenersche Zeitung*, August 24; and Wernicke, "Barackia."

59 **"in front of Landsberger Gate":** Adolf Damaschke, "Die Bedeutung

der Bodenreform für die Moderne Wohnungsnot und den Kleingartenbau," *Gartenflora* 57 (1908): 462.

59 **a phalanx of police and soldiers:** On August 29 demolitions in Barackia, see *Neuer Social-Demokrat*, August 30, 1872, in Geist and Kürvers, *Das Berliner Mietshaus*, 119; and Thomas Lindenberger, "Berliner Unordnung zwischen den Revolutionen," *Strassenpolitik: Zur Sozialgeschichte der öffentlichen Ordnung in Berlin, 1900–1914* (Dietz, 1995), 45–76.

5. THE WORKING-CLASS ORIGINS OF THE GARDEN CITY

63 **regenerated desolate territory:** On farming at Kottbusser Gate, see Hartwig Stein, *Inseln im Häusermeer: Eine Kulturgeschichte des Deutschen Kleingartenwesens bis zum Ende des Zweiten Weltkriegs: Reichsweite Tendenzen und Gross-Hamburger Entwicklung* (Peter Lang, 2000), 238.

64 **"gang of traitors":** Richie, *Faust's Metropolis*, 180.

64 **"invasion of déclassé":** Victor Tissot, *Reportagen aus Bismarcks Reich: Berichte eines Reisenden Franzosen: 1874–1876* (Verl. Neues Leben, 1989).

64 **spontaneous garden shantytowns or arbor colonies:** see Alfred Erlbeck, "Die Sozialhygienische," 40–46; Hartwig Stein, *Inseln im Häusermeer*, 13–19, 280; Heinrich Hinz, *Der Schrebergarten: Praktische Ratschläge zur Einrichtung und Bewirtschaftung von Schreber-, Klein-, und Hausgärten* (Trowitsch & Sohn, 1915).

65 **Workers often rejected them:** Workers disliked their "bourgeois patriarchic, pietistic and nationalist-monarchist aspirations." "Kleingärtnertag!" *Der Kleingärtner*, Nr. 7 (April 2, 1921).

65 **Arbor colonies were different:** For regeneration in colonies, see Kassel and Landgraf, *Zur Geschichte des Organisierten Kleingartenwesens*, 6; Bode Rollka, *Berliner Laubenpieper Kleingärten in d. Grossstadt* (Haude u. Spener, 1987), 32; Max Christian, *Städtische Freiflächen und Familiengärten* (1914), 37.

65 **"Everything belongs in the compost":** "Die Anlegung von Komposthaufen," *Der Kleingärtner* 1 (January 8, 1921): 6. See also, "Wie Steigert Man die Erträge im Gartenbau?" *Der Kleingärtner* 3 (February 1921): 12.

65 **rented on the worst possible terms:** On rental terms, see Johannes Böttner: "Die Laubenkolonien"; and Rollka, *Berliner Laubenpieper*, 33–34; Christian, *Städtische Freiflächen und Familiengärten*, 25–29.

66 **"gulf that separates the workers":** Richie, *Faust's Metropolis*, 179.

66 **Gemütlichkeit (Coziness) I, II, and III:** "Die Ältesten Kleingartenvereine Treptows (1887)," Verband der Gartenfreunde Treptow.

66 **"We eat salad, yes":** Rollka, *Berliner Laubenpieper*, 77.

66 **"We don't want to toil too much":** "Aufgaben und Organisation des Kleingartenamts Berlin," *Der Kleingärtner* 1 (January 8, 1921).

66 **created little worlds within the big city:** On Berlin's social life in garden allotments in 1900, see Rollka, *Berliner Laubenpieper*, 33; Versammlung am January 20, 1917, https://www.kleingaertner-weissensee.de/Dokumente/Dokumente/heft_18.pdf.

67 **Visitors went home with fruits:** From the memories of Frau Dommasch, Rollka, *Berliner Laubenpieper*, 385.

67 **cottages with thin walls:** On property conditions in arbor colonies, see Simon Lengemann, "Kleine Leute in der Großen Depression: Wohnen und Selbsthilfe in Berliner Mietskasernenvierteln und Laubenkolonien 1929–1933," *Jahrbuch Für Die Geschichte Mittel und Ostdeutschlands* 63 (2017): 183–245, 230–2; Vasudevan, *Preoccupations*, 40–41.

67 **voted to create a union:** For information on the union of planters' associations, see Aus Protokollbüchern des Kleingartenvereins "Neu Hoffnungstal," 1914–1953, the Kleingaertner-Weissensee; E. Reinhold, "Die Berliner Laubenkolonisten," (1931) in G. Katsch and J. B. Walz, ed., *Kleingärten und Kleingärtner in 19. und 20. Jahrhundert* (Leipzig, 1996); "Neuorganisation unserer Wirtschaftsgenossenschaft," *Der Kleingärtner* no. 3 (February 1921): xix; and Stein, *Inseln im Häusermeer*, 250.

68 **"grotesque" and unclean:** Alfred Erlbeck, "Die sozialhygienische und Volkswirtschaftliche Bedeutung der Kleingärten für die Städtische Bevölkerung," *Gesundheit* XI, no. 3 (1915): 26–29, 40–43.

68 **"the spirit of community blossoms":** Christian, *Städtische Freiflächen und Familiengärten*, 34.

68 **workers were going to meetings:** On workers organizing in Berlin, see Davis, "'Everyday' Protest and the Culture of Conflict."

70 **failed to conduct a census:** "Abschrift," August 6, 1913, Landesarkhiv Berlin (LAB), Rep 042–08, no. 721; "Der Polizei-Präsident, no. 212, 1.6.12," February 15, 1913, Akten der Abteilung VII, Königlichen Polizei-Präsidiums zu Berlin: 1913–1916, (LAB), A Pr Br. Rep. 030 no. 15890; and District Mayor of Neukölln (name illegible), September 21, 1938, LAB ARep .009, no. 39.

70 **shrubs and vines crawl up:** On photos of Berlin's garden colonies, see Mark Hobbs, "'Farmers on Notice': The Threat Faced by Weimar Berlin's Garden Colonies in the Face of the City's *Neues Bauen* Housing Programme," *Urban History* 39 (May 2012): 263–84.

70 **Howard dreamed up the concept:** For information on Howard's concept of a garden city, see Scott Lloyd and Alexis Kalagas, "Salad Days: Urban Food Futures," *Architectural Design* 91, no. 5 (2021): 40–47.

6. LEAF TOWN

71 **"an idyllic landscape":** Fritz Meissner, *Der Hausgarten* (Ullsein and Co., 1902), 169; and Florian Urban, "The Hut on the Garden Plot: Informal Architecture in Twentieth-Century Berlin," *Journal of the Society of Architectural Historians* 72, no. 2 (2013): 221–49.

71 **There were plenty of conflicts:** On conflicts within arbor colonies, see "Die Laubenkolonien"; and Stein, *Inseln im Häusermeer*, 249.

74 **opted not to accumulate wealth:** On agriculture in Neolithic societies, see Graeber and Wengrow, *Dawn of Everything*, 238, 399; and James C. Scott, *Against the Grain: A Deep History of the Earliest States* (Yale University Press, 2017), chapters 1 and 2.

75 **marvelously productive fens and swamps:** On the productivity of fens and wetlands, see Susan Oosthuizen, *The Anglo-Saxon Fenland* (Windgather Press, 2017), 114–15.

75 **the black earth of the Ukrainian breadbasket:** On black earth as human-engineered, see "The Nebelivka Hypothesis," Forensic Architecture in partnership with David Wengrow, May 18, 2023.

76 **Kropotkin had long been concerned:** On Kropotkin's ecological observations, see Kropotkin, *Fields, Factories and Workshops*; Laura Stark, "Mutual Aid: The Workers' History of Science," *Isis* 114, no. 4 (2023): 841–49.

76 **Parisian farmers were like flies:** For information on Parisian gardening practices, see T. Smith, *A Quarter Acre of French Gardens* (Longmans, Green and Co., 1911).

79 **For decades German chemists tried to draw nitrogen:** For information on the nitrogen cycle, see Hugh S. Gorman, *The Story of N: A Social History of the Nitrogen Cycle and the Challenge of Sustainability* (Rutgers University Press, 2013), 112, 121–22; and Smil, *Enriching the Earth*, 101–3.

79 **"they make the soil themselves":** Kropotkin, *Fields, Factories and Workshops*, 79.

80 **"grow their own food":** Kropotkin, *Fields, Factories and Workshops*, 79.

80 **animal cooperation, not only competition:** For information on Kropotkin's ideas of human cooperation, see Lee Alan Dugatkin, "The Prince of Evolution: Peter Kropotkin's Adventures in Science and Politics," *Scientific American*, September 13, 2011; and Tom Fleischman, *Communist Pigs: An Animal History of East Germany's Rise and Fall* (University of Washington Press, 2020), 123–27.

80 **"200,000 pounds of food":** Kropotkin, *Fields, Factories and Workshops*, 89.

80 **"The Marais system":** G. Stanhill, "An Urban Agro-Ecosystem: The Example of Nineteenth-Century Paris," *Agro-Ecosystems* 3 (1977): 269–84.

7. EVERY WORKING MAN A GARDEN!

82 **free markets took an ugly, unpatriotic turn:** On modern German food and agriculture practices and wartime struggles, see Corinna Treitel, *Eating Nature in Modern Germany: Food, Agriculture and Environment, c.1870 to 2000* (Cambridge University Press, 2017), 163–4; and Alice Weinreb, *Modern Hungers: Food and Power in Twentieth-Century Germany* (Oxford University Press, 2017), 3, 14–15, 17, 20–23, 44–46.

82 **"Spread shit on your bread!":** Belinda Davis, *Home Fires Burning: Food, Politics, and Everyday Life in World War I Berlin* (University of North Carolina Press, 2000), 89.

82 **handed out vacant land for cultivation:** On the increase of gardening in Berlin during World War I, see Tara Zahra, *Against the World: Anti-Globalism and Mass Politics Between the World Wars* (W. W. Norton, 2023), 51; and Nilsen, *Working Man's Green Space*, 90.

83 **They hoarded and gouged:** On leasing companies in Berlin, see "Extract from the Official Stenographic Report," Meeting of the City Council on January 22, 1920, LAB A Rep. 000–02–01, Number: 1865; and "Entstehung der Kleingärten in Berlin und Treptow (1896)," Verband der Gartenfreunde Treptow.

83 **"women of little means":** Davis, *Home Fires Burning*, 76–82.

83 **gardeners won more political backing:** For information on the increased political power of gardeners, see "Abschrift," August 6, 1913, (Bezirksamt Berlin-Steglitz (1920–1945), LAB Rep 042–08, no 72; "Berlin Trade Union Commission to Mr. Chairman of the City Council Dr Weyl," 1920, LAB A Rep. 000–02–01, Number: 1865.

83 **"grow all of the vegetables":** Christian, *Städtische Freiflächen und Familiengärten*, 34–35.

83 **"German People's Park of the Future":** As quoted in Nilsen, *Working Man's Green Space*, 153.

83 **"Every working man a garden!":** in a pamphlet called *Everyman Self-Sufficient!*

84 **Migge called for public ownership:** For information on Migge's agricultural ideas, see David Haney, *When Modern Was Green: Life and Work of Landscape Architect Leberecht Migge* (Routledge, 2010), 104–105, 227; and Leberecht Migge, *Laubenkolonien und Kleingärten* (München, 1917), 12.

85 **"unable to meet its food obligations":** "Proceedings Report of the Committee on the Preliminary Discussion of the Motion of City Councilor Barkowski and Colleges and Dr. Weyl and Colleges on the Establishment of a Municipal Allotment and Settlement Office," Berlin, February 17, 1920, LAB A Rep. 000–02–01, Number: 1865.

85 **demand for allotments grew daily:** On the demand for allotments and the July 1919 Small Gardens Act, see "Extract from the Official Stenographic Report," Meeting of the City Council on January 22, 1920, and Extract from the Official Stenographic Report on the Assembly of the City Council on February 26, 1920, LAB A Rep. 000–02–01, Number: 1865; Kleingarten- und Kleinpachtlandordnung—KGO/ Allotment and Small Holdings Ordinance, July 19, 1919; Nilsen, *Working Man's Green Space*, 96.

85 **7 percent of the city territory:** On Berlin's green spaces after the 1920 expansion, see Barry A. Jackisch, "The Nature of Berlin: Green Space and Visions of a New German Capital, 1900–45," *Central European History* 47, no. 2 (June 2014): 307–33.

86 **"their painful experiences can be included":** Discussed Berlin, April 22, 1920, at the meeting of the City Council, LAB A Rep. 000–02–01, Number: 1865.

86 **"even the simplest worker":** "The Organization of the Settlement and Housing in the New Municipality," July 16, 1920, LAB A Rep. 000–02–01, Number: 1865.

86 **"Give them Gardens!":** "Aufgaben und Organisation des Kleingartenamts Berlin," 3; and "Extract from the Official Stenographic Report," Meeting of the City Council on January 22, 1920, LAB A Rep. 000–02–01, Number: 1865.

86 **"Some daredevils even want":** "Extract from the Official Stenographic Report," Meeting of the City Council on January 22, 1920, LAB A Rep. 000–02–01, Number: 1865.

86 **shield them from the market:** On the need for the state to shield gardeners, "The Organization of the Settlement and Housing in the New Municipality," July 16, 1920, LAB A Rep. 000–02–01, Number: 1865.

86 **finally they rented an allotment:** On Tessen's allotment and the green shantytown, see Urban, "Hut on the Garden Plot."

88 **"were the experts after all":** "Extract from the Official Stenographic Report on the meeting of the City Council on September 30, 1920," LAB A Rep. 000–02–01. Number: 1865.

88 **Once professionals got involved:** On architects raising costs of garden housing out of reach, see "Discussed," Berlin, October 8 and 14, 1920, LAB A Rep. 000–02–01, Number: 1865; and "Extract from the Official Stenographic Report on the Meeting of the City Council on October 19, 1920," LAB A Rep. 000–02–01, Number: 1865.

88 **"everything changes from splendor":** Hobbs, "Farmers on Notice."

89 **"We are constantly monitored":** Versammlung am 29.08.1926, Schriftenreihe zur Geschichte der Weissenseer Kleingartenbewegung,

Informationen Dokumente Analysen, Teil 18, Aus Protokollbüchern des Kleingartenvereins "Neu Hoffnungstal," 1914–1953, Arbeitsgruppe "Weissenseer Kleingärtnertradition), https://www.kleingaertner-weissensee.de/Dokumente/Dokumente/heft_18.pdf.

89 **officials sanctioned developments:** For information on development pressures on arbor colonies, see Hobbs, "Farmers on Notice"; "Demonstration der Kleingärtner in Adlershof," *Südost*, September 5, 1925, in LAB A REP 045–08, nr. 130; and Versammlung am 01.3.1925; 25–26.04.1925, Weissensee Kleingärtner, https://www.kleingaertner-weissensee.de/Dokumente. For large protests against a villa development in arbor colonies in Treptow, see Reichsverband der Kleingartenvereine Deutschlands, Provinzialverband, no date, 1924, LAB A REP 045–08, nr. 130.

89 **"constant burden to the state":** Peter Hall, *Cities of Tomorrow: An Intellectual History of Urban Planning and Design Since 1880* (Wiley-Blackwell, 2014), chapter 2.

90 **"Everyone takes the breeding of horses":** As quoted in Wolfgang Voigt, "The Garden City as Eugenic Utopia," *Planning Perspectives* 4, no. 3 (September 1989): 304.

8. A QUARTER ACRE AND A HOG

93 **sponsored a design competition:** On the Deanwood green housing competition, see Michael Binder, Marcie Meditch, Lael Taylor, and Jenny Wienckowski, "Urban Grapevine: Visions of Regenerative Multifamily Housing in Washington DC," in Ali Sayigh, ed., *Green Buildings and Renewable Energy* (Springer, 2020), 129–41.

94 **"He had a garden . . . So they could prosper":** Alice and Loretta Tate interview part 1, Marshall Heights: Civic Mindedness and Engagement Incarnate, pre-DC Home Rule Oral History Interview, May 26, 2020.

96 **"disgrace to the Capital":** "Discontinuance of Alley Dwellings in D.C.," February 21, 24, 1922, Committee on the District of Columbia, United States Senate, HRG-1922-DCS-0006.

96 **elites in DC went on ghetto tours:** For elite reports on ghetto conditions in Washington, DC, see Charles Frederick Weller and Eugenia Weller, *Neglected Neighbors: Stories of Life in the Alleys, Tenements and Shanties of the National Capital* (John C. Winston, 1909), 212; and Lisa Goff, *Shantytown, USA: Forgotten Landscapes of the Working Poor* (Harvard University Press, 2016), 208.

96 **"the Negro" was poorly rooted:** For a temporal spread of views on "the Negro problem," see Howard Odum, "Social and Mental Traits of the Negro: Research into the Conditions of the Negro Race in South-

ern Towns" (PhD diss., Columbia University, 1910); and Luigi Laurenti, *Property Values and Race: Studies in Seven Cities* (University of California Press, 1960), 10.

96 **officials condemned alley houses:** For information on condemned alley houses, see Weller, *Neglected Neighbors*, 205.

96 **"a large return on the investment":** "Report of Hearings on H.R. 4467," Subcommittee on Education, Labor, and Charities, Committee on the District of Columbia, March 30, 1906, United States House of Representatives, HRG-1906-DCH-0004.

96 **white men marauded in Black neighborhoods:** On mob violence in Black neighborhoods, see Chris Myers Asch and George Derek Musgrove, *Chocolate City: A History of Race and Democracy in the Nation's Capital* (University of North Carolina Press, 2017), 209, 236–37.

97 **"you're just nobody":** Norman E. Dale, Anacostia Oral History Project, 1975, survey instrument 4–197071, tape 13, ACM archives, box 37.

97 **estranged them from traditional foodways:** Natalie Baszile, *We Are Each Other's Harvest: Celebrating African American Farmers, Land, and Legacy* (Amistad, 2021), 10.

97 **Black Washingtonians began to move:** On neighborhoods East of the River, see "War on the Abattoir," *Washington Post*, June 30, 1900; and Alcione M. Amos, *Barry Farm-Hillsdale in Anacostia: A Historic African American Community* (History Press, 2021), 33–38, 57. On the rural quality, see Thomas J. Cantwell, "Anacostia: Strength in Adversity," *Records of the Columbia Historical Society, Washington, D.C.* 49 (1973): 330–70.

98 **first vocational school in the country:** On the National Training School and Deanwood, see Veronica Popp and Danielle Phillips-Cunningham, "Nannie Helen Burroughs and the Descendants of Miriam: Rewriting Nannie Helen Burroughs into First Wave Feminism," *Gender Forum*, no. 79 (2021): 58–136; Emily Marin interview, April 30, 1987, MS2032, series 6, box 19A, folder 15, GW Special Collections; Elizabeth Barker, oral history interviews, 1976–1981 OH-31 Schlesinger Library, Radcliff; and Quentin Allen interview, April 2, 1987, MS2032, series 6, box 19A, folder 5, GW Special Collections.

98 **People fashioned houses with whatever they found:** For information on the do-it-yourself construction of Black neighborhoods in Washington, DC, see "Washington, City and Capital," in *The Negro in Washington* (Federal Writers' Project, Washington, DC, 1937).

98 **"highly cultivated" gardens:** Sandra R. Heard, "Making Slums and Suburbia in Black Washington During the Great Depression," *American Studies* 57, no. 4 (2019): 5–22.

98 **two white real estate developers:** On the financing arrangement with

Howard Gott and Joseph Tepper, see Katherine C. Johnson interview, April 1, 1987, MS2032, series 6, box 19A, folder 12; Turners interview, May 8, 1987, series 6, box 19A; and Mr. and Mrs. Herbert T. Jones interview, May 28, 1987, MS2032, series 6, box 19A, folder 13, GW Special Collections.

99 **"true liberation requires landownership":** Baszile, *We Are Each Other's Harvest*, 10.

99 **"The only way you could survive":** Alexander D. Davis interview, April 13, 1987, MS2032, series 6, box 19a, folder 8, GW Special Collections.

99 **Real estate records show:** For information on real estate records, see Federal Housing Authority, Map of the District of Columbia, 1934, Washingtoniana Collection.

99 **"We planted everything":** Vincent Bunch interview, April 11, 1987, MS2032, series 6, box 19a, folder 7, GW Special Collections.

99 **"We raised hogs and chickens":** Anne H. Oman, "Marshall Heights Settlers Recall the Birth of Their Neighborhood," *Washington Post*, December 15, 1977.

99 **"We had an acre down there":** Ethel G. Greene interview, Anacostia Oral History Project, 1975, ACM. On street vendors, see Ashanté M. Reese, *Black Food Geographies: Race, Self-Reliance, and Food Access in Washington, D.C.* (University of North Carolina Press, 2019), 26–29.

99 **"They got along very well":** McKinley Taylor interview, 1975, Oral History Transcripts, tape 26, ACM.

99 **the US Department of Agriculture calculated:** For information on expenses of Washington backyard gardens in 1917, see H. M. Conolly, "The City and Suburban Vegetable Garden," *Farmers' Bulletin* 936, Washington, DC, February 1918, 10–11.

100 **"We had chickens in our yard":** Turners interview.

100 **"when you say truck farms":** Harold Thompson interview, April 20, 1987, MS2032, series 6, box 19a, folder 18, GW Special Collections.

101 **livestock of Black home gardeners:** For information on Black home gardeners, see 1925 Census of Agriculture (Washington, DC, 1925), 118; Hayden M. Wetzel, *Mangy Curs and Stoned Horses: Animal Control in the District of Columbia from the Beginnings to about 1940* (Creative Commons, Amazon, 2020), 52; and Charles D. Cheek and Amy Friedlander, "Pottery and Pig's Feet: Space, Ethnicity, and Neighborhood in Washington, D.C., 1880–1940," *Historical Archaeology* 24, no. 1 (1990): 34–60.

101 **"They'd work all the summer":** Davis interview.

101 **develop individual and community self-sufficiency:** For information on Black self-sufficiency, see Reese, *Black Food Geographies*, 42–43; and Goff, *Shantytown USA*, 181.

101 **"no streets, no lights, no water":** Oman, "Marshall Heights Settlers Recall."

101 **"about four miles for my man":** "Depression Victims Make Gallant Attempts to Cooperate," *Evening Star*, November 10, 1935, 57.

102 **Black "invasions" of white areas:** Asch and Musgrove, *Chocolate City*, 255.

102 **"The District government . . . a rat terrier does for a mouse":** Alan H. Lessoff, "Washington Under Federal Rule, 1871–1945," in Daum and Mauch, ed., *Berlin–Washington, 1800–2000*, 253.

102 **struggle went on for decades:** For information on the struggles Black residents faced East of the River, see Dorothy M. Roberts interview, April 13, 1987, MS2032, series 6, box 19a, folder 16, GW Special Collections.

102 **"We didn't have garbage trucks":** Norman E. Dale, 1975, and Edith Greene, 1972, Oral Histories, M03–040, box 37 and box 2, Anacostia Community Museum archives.

102 **waste of backyard privies was upcycled:** On waste and animals in urban gardens, see Catherine Brinkley and Domenic Vitiello, "From Farm to Nuisance: Animal Agriculture and the Rise of Planning Regulation," *Journal of Planning History* 13, no. 2 (May 2014): 113–35.

102 **volume of night soil did not fall:** House Committee on the District of Columbia, "Hearing Under H. Res. 154," August 28, 1913, HRG-1913-DCH-0043; and "Budget Requirements of the District of Columbia," Hearings before the Joint Subcommittee on Fiscal Affairs, March 1947, 331.

102 **"If I wanted to redeem":** H. W. Wiley and Ervin E. Ewell, "The Fertilizing Value of Street Sweepings," *Bulletin* no. 55 (U.S. Department of Agriculture, 1898).

102 **houses had outdoor privies:** For information on sanitation and sewage in Black neighborhoods, see "Conference on Marshall Heights: Hearings before a Subcommittee . . . Eighty-First Congress, First Session, December 12 and 13, 1949"; "Transportation of Refuse in the District of Columbia," Hearing before a Subcommittee of the Committee on the District of Columbia, June 14, 1921, H.R. 5767; House Committee on the District of Columbia, "Hearing Under H. Res. 154," August 28, 1913, HRG-1913-DCH-0043; and "Law Barring the Removal of DC Garbage to be Tested," *Washington Times*, May 21, 1917, 1.

104 **major architects of human life:** For information on the loss of diversity in the human microbiome, see Emmanuelle Le Chatelier et al., "Richness of Human Gut Microbiome Correlates with Metabolic Markers," *Nature* 500 (August 2013), 541–46.

104 **Streams flowed above ground:** "Flood in Suburbs Forces Hundreds from their Beds," *Evening Star*, July 13, 1922.

104 **Elders recalled the groves:** For oral interviews on life in Black neighborhoods East of the River, see Phillip Gregory and Vincent Bunch interview, April 1, 1987, and April 11, 1987, MS2032, series 6, box 19a, folder 7, GW Special Collections; and Pierre McKinley Taylor interview and George Trivers, 1975, Oral History Transcripts, ACM.

105 **also left room for other species to prosper:** On regenerative farming, see Liz Carlisle, *Healing Grounds: Climate, Justice, and the Deep Roots of Regenerative Farming* (Island Press, 2022), 123.

105 **pieced together alternative infrastructure:** On mutual aid associations of the East River, see Phillip Gregory interview, April 1, 1987, MS2032, series 6, box 19a, folder 11, GW Special Collections; "Ask for a Primary School," August 26, 1906, 2; "Board of Education Head Declares Lack of Proper Inspection," *Washington Post*, October 15, 1908, 2; and "Pupils Kept from School Face Law," *Evening Star*, September 20, 1920.

105 **slaughter the fattened hogs:** Loretta Tate, June 2021 Oral History Coffee Chat, YouTube.

105 **"a co-op type of thing":** Norman E. Dale, Anacostia Oral History Project, 1975, survey instrument 4–197071, tape 13, box 37, ACM archives.

105 **"put the pig on a spit":** Alice and Loretta Tate interview part 1.

106 **how people make kin:** For information on the social microbiome and making kin, see Amar Sarkar et al., "Microbial Transmission in the Social Microbiome and Host Health and Disease," *Cell* 187, no. 1 (January 2024): 17; and Donna J. Haraway, "Making Kin in the Chthulucene: Reproducing Multispecies Justice," in A. E. Clarke and D. Haraway, ed., *Making Kin Not Population* (Prickly Paradigm Press, 2018).

9. SHANTYTOWN IN THE WOODS

107 **"raise a part or all of their food":** "US Department of Interior to Dr. Frank App," December 1, 1933, National Archives Record Administration (NARA) RG 96, PI 118, entry 9, box 1.

107 **set up the Division of Subsistence Homesteads:** For information on the homestead subsistence division, see "Project Registration and Examination Recorded by Interview"; Robert M. Carriker, *Urban Farming in the West: A New Deal Experiment in Subsistence Homesteads* (University of Arizona Press, 2010), 4; and Donald Holley, "The Negro in the New Deal Resettlement Program," *Agricultural History* 45, no. 3 (July 1971): 181–83.

108 **vowed to eliminate all alley housing:** Asch and Musgrove, *Chocolate City*, 254.

109 **"We built a place big enough":** Turners interview.

109 **"We plan to have our own cooperative stores":** Florence M. Collins, "Marshall Heights Folk Wary of Any Commercial Invasion," *Afro-American*, March 2, 1935, 13.

109 **Midwives delivered babies:** Debbie and Charles Thomas interview, September 28, 2020, Marshall Heights: Civic Mindedness and Engagement Incarnate, pre-DC Home Rule Oral History Interview.

109 **mutual aid formalized into cooperative movements:** On cooperative movements in Washington, DC, see Collins, "Marshall Heights Men's Self-Help Project Dies, but Women's Flourishes," *Afro-American*, March 9, 1935, 1; and "Subsistence Gardens under Chest Control," *Sunday Star*, June 11, 1939, C-6.

110 **Black Americans created cooperative garden neighborhoods:** For information on cooperative garden neighborhoods around the country, see Andrew Wiese, *Places of Their Own: African American Suburbanization in the Twentieth Century* (University of Chicago Press, 2004); Joseph Stanton Cialdella, *Motor City Green: A Century of Landscapes and Environmentalism in Detroit* (University of Pittsburgh Press, 2020), 32–43; and Gilberto Mazzoli, "Portable Natures: Environmental Visions, Urban Practices, Migratory Flows: Agriculture and the Italian Experience in North American Cities, 1880–1940" (PhD diss., European University Institute, 2023).

110 **"Your neighbor may have tomatoes and squash":** Barry Yeoman, "The Hidden Resilience of 'Food Desert' Neighborhoods." *Barry Yeoman* (blog), August 30, 2018.

112 **"Anyone could pick":** Author interview with Bishop Jeremiah Murphy and Joyce Cox, Marshall Heights, March 24, 2024.

112 **"We're friends out here!":** as cited in Cialdella, *Motor City Green*, 39–40.

112 **"We knew nothing about politics":** George Trivers; Pierre McKinley and Sarah Taylor interview, Anacostia Oral History Project, 1975, Oral History Transcripts, ACM, tape 26.

112 **contemporary political voices in the Black community:** For information on mutual aid and cooperatives in Black political thought, see Jessica Gordon Nembhard, *Collective Courage: A History of African American Cooperative Economic Thought and Practice* (Pennsylvania State University Press, 2014), 148–54; Monica M. White, *Freedom Farmers: Agricultural Resistance and the Black Freedom Movement* (University of North Carolina Press, 2019), 54; and Asch and Musgrove, *Chocolate City*, 210–12.

113 **"will stimulate and for a while produce good results":** White, *Freedom Farmers*, 48.

113 **"A walk through the woods":** Carlisle, *Healing Grounds*, 79.

113 **sustainable self-sufficiency promoted a quiet flourishing:** For

information on Black American spaces of self-determination, see Rebecca J. Scott and Michael Zeuske, "Property in Writing, Property on the Ground: Pigs, Horses, Land, and Citizenship in the Aftermath of Slavery, Cuba, 1880–1909," *Comparative Studies in Society and History* 44, no. 4 (2002): 669–99.

114 **fault lines ran through Black communities:** For information on divisions within Black communities East of the River, see Anita Blake and Irene Donnelly interview, 1975, Oral History Transcripts, tape 29; and Heard, "Making Slums and Suburbia."

114 **"These people in Marshall Heights":** "Alley Homes Were Better, She Avers," *Washington Times*, February 4, 1935, 1.

114 **"a real love of country here":** "Depression Victims Make Gallant Attempts to Cooperate," *Evening Star*, November 10, 1935, 57.

114 **prejudicial banking kept . . . lagging behind:** For information on African American property ownership, see Keeanga-Yamahtta Taylor, *Race for Profit: How Banks and the Real Estate Industry Undermined Black Homeownership* (University of North Carolina Press, 2019), 31.

115 **"a thriving neighborhood":** Collins, "Marshall Heights Folk."

115 **"If you left Deanwood":** Alexander D. Davis interview, April 13, 1987, MS2032, series 6, box 19a, folder 8, GW Special Collections.

115 **What shines through:** For more stories of African American domestic life, see Elizabeth Clark-Lewis, *Living In, Living Out: African American Domestics and the Great Migration* (Kodansha International, 1996), 155–160.

115 **"universe of obligation":** Helen Fein, *Accounting for Genocide: National Responses and Jewish Victimization During the Holocaust* (Free Press, 1979), 4.

115 **The DC census shows:** For more census information, see 1950 US Census, DC, track 93 for Marshall Heights and for Deanwood Federal Housing Administration, "Map of the District of Columbia," DC Public Library, The People's Archive, Washingtoniana Map Collection; and census tract 78, sixteenth census of the United States, District of Columbia, 1940. In Marshall Heights, District Committee surveys reported that two-thirds to 90 percent owned their own homes: "Conference on Marshall Heights: Hearings before a Subcommittee . . . Eighty-First Congress, First Session, December 12 and 13, 1949."

115 **"You couldn't *buy* a job":** Turners interview.

10. PURIFYING THE WARTIME GARDEN

117 **politics of "blood and soil":** "Statement on the Small Garden Association," January 1, 1934, LAB, A Pr Br, Rep. 030–04, Nr. 4345. For hopes of doubling the number of Berlin's plots, see "New Organization of the Small Gardens," *Spandau Zeitung*, January 11, 1939, nr. 29.

118 **"Germany will live":** "Weissenseer Kleingärtner im 3. Reich," Weissenseer Kleingärtnertradition.

118 **"degenerative and sick":** Treitel, *Eating Nature in Modern Germany*, 159–75.

118 **In botany courses soldiers learned:** On botany in Nazi Germany, see Alhena Katsof, "Mythological Formulations: Autochthony at the Root of the Matter," in Gerard Forde et al., *On the Necessity of Gardening: An ABC of Art, Botany and Cultivation* (Valiz, 2021), 23–26.

118 **projects of plant, animal, and human breeding:** On back breeding and agricultural experiments in Nazi Germany, see Diane Ackerman, *The Zookeeper's Wife* (W. W. Norton, 2007), chapter 8; and Tiago Saraiva, *Fascist Pigs: Technoscientific Organisms and the History of Fascism* (MIT Press, 2016), 5–13.

118 **Zygmunt Bauman suggested that the Holocaust:** Zygmunt Bauman, *Modernity and the Holocaust* (Polity Press, 1989), 13.

119 **"Negro villages":** Urban, "Hut on the Garden Plot."

120 **"were all connected":** Böhmer, Krüger, Arndt, "Strafsache gegen . . . ," March 10, 1933, LAB B Rep. 031–02–01, Nr. 12698.

120 **"little Moscows":** Bernd Hildebrandt, *Abschied von Der Laube: Die Zeit vor der Entstehung des Märkischen Viertels: Bürger Erforschen Ihren Ortsteil* (G. Grosskopf, 1988), 83.

120 **instigated a purge of allotments:** On the purge of garden allotments by Nazis, see "Weissenseer Kleingärtner im 3 Reich," Weissenseer Kleingärtnertradition; "To the Mayo or Halle," January 2, 1936; "Office of Statistics to Office of City Planning [Berlin]," October 11, 1937; "District Mayor of Neukölln (name illegible), September 21, 1938, LAB A Rep .009, Nr. 39; and "Entwurf einer Pressenotiz," January 30, 1936, LAB Rep 042–08, no 721.

120 **"low-quality stock":** Albert Schorr, "Die Wohnlauben in Berlin," *Der Gemeindetag Zeitschrift* 25 (1935): 793–97.

121 **"gained meaning again":** "Entwurf einer Pressenotiz Der Zuzug in Lauben auf Pachtland polizeilich verboten!" January 30, 1936, LAB Rep 042–08, no 721.

121 **safe havens for political activism:** For information on arbor colonies as political safe havens, see "Weissenseer Kleingärtner im 3. Reich"; and Catharina Paetzelt, *Die Ersten 100 Jahre*, 80.

122 **"decrees of the building police":** "Warnung vor eigenmächtigen Bauen, besonders in Laubengebieten," March 25, 1937, LAB Rep 042-08, no 721.

122 **"Gardeners in no way":** "Entwurf einer Pressenotiz," January 30, 1936, LAB Rep 042-08, no 721.

122 **"who were true Germans":** Lengemann, "Kleine Leute in der Großen Depression," 230.

122 **"sewage farm":** Gray, "Urban Sewage and Green Meadows."

122 **"Berlin could colonize itself":** "Kläranlagen oder Rieselfelder?" *Städtischer Nachrichtendienst*, October 21, 1930.

122 **gated garden community in the city of Bremen:** On the Bremen project, see Voigt, "Garden City as Eugenic Utopia," 295–312.

123 **informal settlements grew in number:** On residency in Berlin garden houses, see "Warnung vor Eigenmächtigen Bauen, Besonders in Laubengebieten," March 25, 1937, LAB Rep 042–08, no. 721; and "Warnung der baupolizeilichen," January 30, 1937, LAB Rep 042–08, no. 721.

123 **"battle for production":** On gardens and the battle for production, see "Tätigkeitsbericht, 1933–1937, Provinzgruppe Berlin, 1938, LAB, A Pr Br, Rep. 030–04, Nr. 4345; and Hildebrandt, *Abschied von Der Laube*, 119.

124 **sheltered in arbor colonies:** On Jews sheltering in arbor colonies, see Paetzelt, *Die Ersten 100 Jahre*, 80; and Hildebrandt, *Abschied von Der Lauben*, 108–09.

124 **Bombing raids drove middle-class Berliners:** On dwellings during bombings, see Hilbrandt, *Housing in the Margins*, 40; and Urban, "Hut on the Garden Plot."

125 **East Germans relied on their plots:** On gardens in East Berlin, see Moray McGowan, "Playing Leapfrog on the People's Potatoes: The Role of the *Kleingarten* in the GDR," in Graham Bartram and Anthony Waine, eds., *Culture and Society in the GDR* (University of California, 1984), 71–82.

125 **"cherries for higher prices":** Author interview with Arbeitsgruppe Weissenseer Kleingärtnertradition, June 4, 2023.

125 **His garden fed his family:** For information on self-provisioning in East Germany, see Fleischman, *Communist Pigs*, 122–27.

126 **"rape of the land":** Sarah Ruth Wilson, "'Rape of the Land': 21st Century Ecofeminism and Environmental Rape Culture," *Oak Tree Notebook*, February 24, 2016.

128 **Specialized bacteria ally with root tips:** For information on beneficial bacteria in soils, see S. N. Pandey et al., "Classification of Soil Microorganisms," in Dilfuza Egamberdieva and Parvaiz Ahmad, eds., *Plant Microbiome: Stress Response* (Springer, 2018), 1–19; and Anna Katrina Hunter, "Microbes in Soil Help Sorghum Stay Strong Against Droughts," *Inside Science*, April 17, 2018.

128 **cleaner than the usual "biosolid":** This was before the discovery of the PFAS toxin in agricultural fields from biosolid use. For more information, see USDA, "Per- and Polyfluoroalkyl Substances FAQ,"

Farmers.gov, January 24, 2024. https://www.farmers.gov/protection-recovery/pfas/faq.

130 **gardens on the urban perimeter of Chinese cities:** On urban gardens in Beijing, see Joshua Goldstein, *Remains of the Everyday: A Century of Recycling in Beijing* (University of California Press, 2021), 34–36.

131 **"birds singing on the bushes":** Charles Darwin, *On the Origin of Species by Means of Natural Selection* (William Collins, 2012), 489.

11. THE WAR AT HOME

134 **The Victory Gardens were phenomenally fertile:** On Washington's Victory Gardens, see Lessoff, "Washington Under Federal Rule," 235–62; "Front-Yard Garden Ban Is Lifted by Commissioners." *Washington Post*, May 11, 1943, 1; and Judith Sumner, *Plants Go to War: A Botanical History of World War II* (McFarland & Company, 2019), 7.

134 **"unfinished city":** Daniel Patrick Moynihan, "Employment, Income, and the Ordeal of the Negro Family," *Daedalus* 94, no. 4 (1965): 745–70.

134 **"The present population is scattered":** "Conference on Marshall Heights," 27; Hearings before a Subcommittee, Eighty-First Congress, First Session, December 12 and 13, 1949: 27.

134 **carry out the job of "slum clearance":** For information on the "slum clearance" of Marshall Heights, see "Conference on Marshall Heights," 29; "History of Negro Subdivision," *Evening Star*, February 1, 1944, 7; Mary Spargo, "Slum Clearance," *Washington Post*, September 8, 1943; and Asch and Musgrove, *Chocolate City*, 323–24.

135 **"No. 1 project":** "Planners K. O. Business Groups," *Washington Daily News*, January 27, 1949.

135 **"opportunity for a laboratory experiment . . . a rare chance to study":** Ben McKelway, "A Possible Experiment at Marshall Heights," *Evening Star*, March 18, 1945, 46.

135 **"President U. S. Grant was very kind":** Author interview with Joyce Cox, Washington, DC, March 23, 2024.

135 **"a slum as evil":** Murrey Marder, "Segregation Report Assailed as Distorted 'Special Plea,'" *Washington Post*, December 22, 1948.

135 **"District of Co-Slum-Bia":** "Conference on Marshall Heights," 53.

136 **Marshall Heights residents fought:** For defense of Marshall Heights, see testimony of Patsy Allen and Howard D. Woodson, "Conference on Marshall Heights," 95–98; and Luther P. Jackson, "Marshall Heights View Offers," *Washington Post*, November 18, 1961.

136 **"swept away":** "Conference on Marshall Heights," 39, 53.

136 **neglected to install basic infrastructure:** On Marshall Heights con-

ditions and infrastructure, see "Urge Further Study on Marshall Heights," *Home Builders Monthly*, April 1949; and "Conference on Marshall Heights," 29, 56.

136 **"health menace":** "New Appeal Is Made for Slum Funds," *Washington Post*, March 6, 1948.

137 **"comprehensive plan":** Robert Taylor, "Marshall Heights Residents Again to Fight for Homes, *Pittsburgh Courier*, February 19, 1949.

137 **"overflow from congested areas":** "Conference on Marshall Heights," 29, 56.

137 **"forty others can be housed":** "Conference on Marshall Heights," 84.

137 **"crush the Negro":** "Marshall Heights Citizens Unit Opposes Redevelopment Plans," *Times Herald*, May 29, 1949.

137 **"a Hitler-like ghetto":** Dorothea Andrews, "Slum Plans Spelled Out for 2 Areas in NE, SE" *Washington Post*, March 3, 1948.

137 **"the logical dividing line":** Murrey Marder, "Segregation Report Assailed as Distorted 'Special Plea,'" *Washington Post*, December 22, 1948.

138 **jobs designated for Black people:** On jobs for Black labor, see Heard, "Making Slums and Suburbia," note 66.

138 **"hydraulic societies":** Karl Wittfogel, *Oriental Despotism: A Comparative Study of Total Power* (Yale University Press, 1957), 9–10.

138 **"water pollution threat":** Sam Zagoria, "D.C. Fails to End Water Peril After 56 Months," *Washington Post*, January 21, 1951.

138 **Privies have a poor reputation:** For information on privies, see Sarah Newman, *Unmaking Waste: New Histories of Old Things* (De Gruyter, 2023), 21.

139 **ecological dead zone:** For information on the Anacostia River and contamination, see "Outdoor Plumbing Ban Sought by Cary to End Water Pollution Threat," *Evening Star*, September 10, 1949, 34. On Fort Dupont Reservoir, see Cantwell, "Anacostia," 333; and Abigail Ulman, "Expanding Conceptions of Water Security and Health: A Case Study in Anacostia, D.C." (thesis, Walsh School of Foreign Service, Georgetown University, 2019).

139 **as racialized as other financial tools:** On unequal municipal bond financing, see Destin Jenkins, "How to Achieve Systemic Equality," 2023, YouTube.

140 **"reduces the cost":** "Conference on Marshall Heights," 51.

140 **overestimated the cost of infrastructure:** Commissioners spent $100,000 installing waterlines, not the projected two million dollars. See "Marshall Heights Is Still a Problem," *Washington Daily News*, February 12, 1952.

140 **Marshall Heights was not blighted:** On overcalculating blight in Marshall Heights, see "Conference on Marshall Heights," 96.

140 **dozens of photos of the neighborhood:** For John Wymer's photos of Marshall Heights, see John P. Wymer photograph collection, DC History Site.

141 **"I think you ought to read the records very carefully":** "Conference on Marshall Heights," 89–90.

142 **"You will not want to live":** "Conference on Marshall Heights," 95.

142 **neighborhoods had already campaigned:** On campaigns against low-income housing, see "Rezoning Plan for Far Northeast Given Study," *Evening Star*, January 7, 1944, 20.

142 **"If people are moved out":** "Conference on Marshall Heights," 99–100.

142 **"Thousands of Negroes":** "Marshall Heights Citizens Unit Opposes Redevelopment Plans," *Times Herald*, May 29, 1949.

142 **"Pilgrims who landed on Plymouth Rock":** "Conference on Marshall Heights," 90.

142 **"When my husband and I":** "Conference on Marshall Heights," 90.

142 **"Black ghetto":** Martha Strayer, *Washington Daily News*, February 3, 1950, VF.001, Washingtonia Collection.

142 **"slum work":** Thomas Winship, "Officials Agree to Call Off Marshall Heights Slum Work," *Washington Post*, March 23, 1950.

143 **"national figure in slum clearance":** "FHA Press Release," May 2, 1953, RGI-s2-b16-FightBlight, Columbia Maryland Archives (CMA).

143 **Developers understood "blight" literally:** For configuration of neighborhoods as blight, see "General Remarks, Rehabilitation of Residential Properties," July 14, 1953, RGI-s2-b16-fFightBlight, CMA.

144 **"law enforcement machinery equipped with a housing code":** James Rouse to Guy T. Hollyday, Commissioner Federal Housing Administration, August 18, 1953, RGI-s2-b16, Fight Blight, CMA.

144 **new infrastructure turned out to be punitive:** On punitive infrastructure in Marshall Heights, see "DC Kills $8,916 Assessment in Marshall Heights," *Evening Star*, December 5, 1956.

144 **"The financial strain is terrific":** "Marshall Heights Is Still a Problem," *Washington Daily News*, February 12, 1952.

144 **Inspectors descended on Marshall Heights:** For information on building inspectors in Marshall Heights, see Luther P. Jackson, "Marshall Heights Spurns Talk of Redevelopment," *Washington Post*, November 21, 1960. For the map, see James W. Rouse and Nathaniel S. Keith, *No Slums in Ten Years: A Workable Program for Urban Renewal* (US Government Printing Office, 1955), 27.

144 **"a grave health menace":** Donald B. Hadley, "Donohue Assails Marshall Heights Shocking Homes," *Evening Star*, October 17, 1952, 1; and

Paul Sampson, "Marshall Hgts. Renewal Reported as Advancing," *Washington Post*, October 15, 1957.

144 **people dumped old cars:** On Marshall Heights as dumping ground, see Luther P. Jackson, "Marshall Heights View Offers," *Washington Post*, November 18, 1961.

144 **pulling down 7.5 million dwelling units:** On demolitions, see Francesca Russello Ammon, *Bulldozer: Demolition and Clearance of the Postwar Landscape* (Yale University Press, 2016), 5.

145 **junkyard for unwanted citizens:** On urban renewal in Southeast DC, see Asch and Musgrove, *Chocolate City*, 321–25, 331, 348, 363.

146 **unemployment . . . reached 50 percent:** See Derek Musgrove, "Black Power in Washington, DC, 1968–1991," Black Power in Washington, DC, interactive map.

146 **short life expectancies:** For information on short life expectancy in Anacostia, see Ulman, "Expanding Conceptions of Water Security and Health," 39.

147 **"discrimination in the provision of municipal services":** "Mary Burner et al. vs. Walter E. Washington," US District Court, DC, July 15, 1975.

147 **mailed homeowners bills for $900:** On overcharging of Black residents, see Bill Allegar, "City Repair Bills Irk Marshall Heights," *Washington Times*, October 3, 1985; and Taylor, *Race for Profit*.

147 **Property . . . mattered deeply:** On the Marshall Heights Community Development Corporation, see Susie McFadden-Resper and Brett Williams, "Washington's 'People without History,'" *Transforming Anthropology* 13, no. 1 (April 2005): 9.

147 **"keep that house or shanty":** Oman, "Marshall Heights Settlers Recall."

147 **"Own it, build it":** Baszile, *We Are Each Other's Harvest*, 10.

147 **"to sell it to white people":** Norman E. Dale, 1975, survey instrument 4–197071, tape 13, box 37, ACM archives.

12. GREEN PRIVILEGE

152 **"My yard is agriculture":** Georgette Norman, *Montgomery Advertiser*, July 12, 1998, 13A.

152 **"If you live in a neighborhood":** John F. Maclean, "City Launches Latest Attack in War against Weeds," *Montgomery Advertiser*, January 22, 1999, C1.

152 **"of no value":** City of Montgomery v. Norman, Court of Criminal Appeals of Alabama. CR-98–0837. Decided: August 27, 1999.

152 **"The zoning inspector":** Author telephone interview with Georgette Norman, July 6 and July 7, 2021.

152 **"I told her she could take":** Amy Frasier, "Appeals Court Reverses Lower Ruling on Woman's Lawn," Associated Press, August 27, 1999.

153 **"They convicted me in criminal court":** Author telephone interview with Georgette Norman, July 6, 2021.

153 **West Montgomery had a park:** For information on conditions in West Montgomery, see United States, "U.S. Censuses of Population and Housing: 1960. Census Tracts. Final Report PHC(1)-1-180," Superintendent of Documents, US Government Printing Office, 1961.

154 **"Although Norman may have intended":** City of Montgomery v. Norman.

155 **"My neighborhood was plagued with car-based prostitution":** City of Montgomery v. Norman.

156 **"neatly and orderly maintained":** In 1991, the city liberalized the ordinance, see "An Ordinance No. 63–62," *Alabama Journal*, December 13, 1962, 8-D.

156 **Turf grass is a product of major marketing:** For information on turf grass and lawns, see Sarah B. Schindler, "Banning Lawns," *George Washington Law Review* 82, no. 2 (2014): 394–455; Virginia Scott Jenkins, *The Lawn: A History of an American Obsession* (Smithsonian Books, 1994); Ted Steinberg, *American Green: The Obsessive Quest for the Perfect Lawn* (W. W. Norton, 2006); Paul Robbins, *Lawn People: How Grass, Weeds, and Chemicals Make Us Who We Are* (Temple University Press, 2007); and Michael Pollan, "Why Mow? The Case Against Lawns," *New York Times*, May 28, 1989.

158 **lawns served as an ideology:** On lawns as ideology, see Robbins, *Lawn People*, 27.

158 **"Shiftless home builders":** Jenkins, *Lawn*, Kindle loc. 1259.

158 **"chicken yards" into lawns:** Sylvia Hood Washington, "Mrs. Block Beautiful: African American Women and the Birth of the Urban Conservation Movement, Chicago, Illinois, 1917–1954." *Environmental Justice* 1, no. 1 (2008): 13–23.

159 **"It's something of an achievement":** Amanda I. Seligman, *Chicago's Block Clubs: How Neighbors Shape the City* (University of Chicago Press, 2016), 134.

159 **"Nobody cared what we wanted' ":** Jane Jacobs, *The Death and Life of Great American Cities* (Random House, 1961), 15.

159 **"a wasteland":** Adam Rome, *The Bulldozer in the Countryside: Suburban Sprawl and the Rise of American Environmentalism* (Cambridge University Press, 2001), 122.

160 **Landowners "treat" their lawns:** On lawn treatment chemicals, see Robbins, *Lawn People*, 64.

13. THE VEGETAL LINE

164 **white vigilantes attacked Black neighbors:** On crossing the color line, see Beryl Satter, *Family Properties: Race, Real Estate, and the Exploitation of Black Urban America* (Metropolitan Books, 2009); and Washington, "Mrs. Block Beautiful."

164 **"Mob action and violence":** quoted in Seligman, *Chicago's Block Clubs*, 45.

165 **Yard care ordinances worked in tandem:** On yard care ordinances, see Bret Rappaport, "Green Landscaping: Greenacres," *John Marshall Law Review* 26/4 (1993).

165 **to sit on a wall:** "Vags Must Not Loaf on Walls, Say Police," *Montgomery Advertiser*, September 13, 1907.

166 **fines for yards in the city of Chicago:** On weed fines in Chicago, see Chris Coffey, "Gardeners Challenge Chicago Weed Control Rules," *NBC Chicago News*, July 31, 2014.

166 **The growth of homeowners associations:** On HOAs, see Courtney Ruby, "Let It Grow: Freeing the Lawn from Aesthetically Rigid and Environmentally Damaging Real Covenants," *UMKC Law Review* 87 (2018): 435.

166 **"Careless, sloppy people":** J. E. Mitchell, "Camarillo Will Consider Strict Landscape Measures," *Los Angeles Times*, December 13, 1994.

167 **White gardeners caught in violation:** For stories of green privilege, see Story Hinckley, "Is There a War on Front Yard Gardens?" *Christian Science Monitor*, June 9, 2016; Elaine Viets, "City Garden Nipped in the Bud," *St. Louis Post Dispatch*, August 11, 1994, 03G; Tom Christopher, "Vegetables Can Enliven Front Yard," *St. Louis Post Dispatch*, May 4, 1995, 02; "Town Outlaws Gardens in Front Yards," *Newstex Trade & Industry Blogs*, Newstex, June 9, 2016; "Neighborhood Dispute over Unmanicured Yard Headed to Court," January 3, 2000, D7; and "Neighbors Sue over High Grass," *Columbus Dispatch*, Oct 19, 2000, 4C.

167 **"Gardening While Black" hashtag:** Janelle Griffith, "Detroit Man Sues Women Who Allegedly Called Police on Him for 'Gardening While Black,'" NBC News, March 5, 2019; and author telephone interview, Denise Morrison, June 22, 2021.

168 **"not necessarily familiar faces":** City of Falcon Heights Regular Meeting of the City Council, City Hall, 2077 West Larpenteur Avenue, May 27, 2020, 7:00 p.m.

168 **"I feel targeted":** City of Falcon Heights Regular Meeting, May 27, 2020.

168 **The more homogeneous the social fabric of a community:** On the lack of biodiversity in homogeneous communities, see James Curtis Fraser et al., "Covenants, Cohesion, and Community: The Effects of Neighborhood Governance on Lawn Fertilization." *Landscape and Urban Planning* 115 (July 2013): 30–38.

14. HOOP DREAMS

174 **as long as the hoop house was temporary:** Elmhurst City Council Meeting, January 1, 2017.

174 **The neighbor was worried:** Author telephone interview with neighbor who wished to remain unnamed.

174 **both a temporary and a permanent structure:** Administrative Hearing, Re: City of Elmhurst, Petitioner vs. Nicole Virgil and Dan Virgil, Respondents, Citation No. EL002450, transcription from proceedings before Hearing Officer Jeffrey Greenspan, January, 24, 2016.

175 **Americans divide land by use:** On US zoning and land use, see Sonia A. Hirt, *Zoned in the USA: The Origins and Implications of American Land-Use Regulation* (Cornell University Press, 2014), 31–50.

175 **"a subsistence farm?":** Elmhurst City Council Meeting, February 4, 2019.

176 **"As in all Utopias":** Jacobs, *Death and Life of the American City*, 17.

176 **"We are a sustainable community":** Elmhurst City Council Meeting, February 4, 2019.

176 **"This is a suburban setting":** Elmhurst City Council Meeting, February 4, 2019.

177 **"I am certainly conspicuous":** Author telephone interview with Nicole Virgil, June 21, 2021.

177 **"What a lot of these restrictions are designed to do":** Author telephone interview with Isaiah Jeremie and Ari Bargil, April 1, 2021.

177 **"cultivate vegetable gardens":** "Vegetable Garden Protection," Illinois HB0633, effective January 2022.

177 **Illinois joined Iowa, Florida, and Maryland:** For information on Maryland landscaping bill, see Maryland HB322, House Bill, "Low-Impact Landscaping."

178 **battle against hoop houses:** For more on Elmhurst's battle against hoop houses, see David Giuliani, "Long Battle Over?" *Patch*, June 14, 2022, https://patch.com/illinois/elmhurst/long-battle-over-elmhurst-panel-wants-tent-ban-lifted.

15. SUPERPOWER SELF-PROVISIONING

181 **Estonian CO_2 emissions dropped:** On CO_2 emissions in the former Soviet Union, see Janis Brizga et al., "Drivers of CO_2 Emissions in the Former Soviet Union," *Energy* 59 (September 2013): 743–53.

182 **thousands of people toiling in the earth:** On Suur-Sõjamäe gardens, see Aljona Surzhikova, film, "Not My Land," (Tallinn, 2013); and Sonia A. Hirt, *Iron Curtains: Gates, Suburbs and Privatization of Space in the Post-Socialist City* (Wiley-Blackwell, 2012).

183 **his plot and hut, called in Russian, a *dacha*:** On dacha gardens in post-Soviet Russia, see Caleb Southworth, "The Dacha Debate: Household Agriculture and Labor Markets in Post-Socialist Russia," *Rural Sociology* 71, no. 3 (September 2006): 451–78.

184 **Soviet citizens won the right to garden:** For a good review of legislation, see Aleksandra K. Kasatkina, "Dachnye Razgovory kak ob'ekt etnograficheskogo issledovaniia: razrabotka metoda" (PhD diss., Musei antropologii i ethnografii im, Petra Velikogo, St. Petersburg, 2019), 61–64.

184 **"Each factory, each city":** William Moskoff, *The Bread of Affliction: The Food Supply in the USSR during World War II* (Cambridge University Press, 1990), 963.

184 **By 1944, sixteen million urbanites dug:** For information on urban farming and dacha communities in Soviet cities, see Stefan Hedlund, *Private Agriculture in the Soviet Union* (Routledge, 1989), 35; Tiina Tammet, "From Post-War Gardening Plots to Cooperative Holiday Villages," in Epp Lankots and Triin Ojari, eds., *Leisure Spaces: Holidays and Architecture in 20th Century Estonia* (Estonian Museum of Architecture, 2020), 120–35; Andrei Markevich, "Finding Additional Income: Subsidiary Agriculture in Soviet Urban Households, 1941–1964," in Donald Filtzer et al., eds., *A Dream Deferred: New Studies in Russian and Soviet Labour History* (Peter Lang, 2008), 386–414; and Diana Mincyte, "Everyday Environmentalism: The Practice, Politics, and Nature of Subsidiary Farming in Stalin's Lithuania," *Slavic Review* 68, no. 1 (2009): 31–49.

185 **The only dip:** For information on urban self-provisioning under Khrushchev, see Markevich, "Finding Additional Income," figure 16.3.

186 **"when we first saw it":** Author interview with Tiuu Saarist, Paide, Estonia, September 10, 2021. My thanks to Epp Lankots for interpreting.

186 **make sure gardeners complied with the rules:** For information on Soviet gardening policies, see Naomi Roslyn Galtz, "The Strength of Small Freedoms," in Daniel Bertaux, Paul Thompson, and Anna Rotkirch, eds., *Living Through the Soviet System* (Transaction Publishers, 2004): 176–92.

187 **"Industrial Man to Hunter-Gatherer":** Alan Philips, "Russians Search for Food in Forests: Passive and Leaderless, Their Thoughts Are Far from Revolution," *Daily Telegraph/Edmonton Journal*, September 8, 1998, A4.

187 **failed subjects of a failed modernization project:** On Soviet self-provisioning and modernization, see Clifford G. Gaddy and Barry W. Ickes, *Russia's Virtual Economy* (Brookings Institution Press, 2002), 168.

187 **they moonlighted to design playful, ecologically thoughtful:** On dacha houses, see Lankots and Ojari, eds., *Leisure Spaces*, 57–58.

187 **Soviet urban farmers:** On Soviet urban farming, see Louiza M. Boukharaeva and Marcel Marloie, *Family Urban Agriculture in Russia: Lessons and Prospects* (Springer, 2015), 24, 66.

188 **Ivan Michurin crossed an ornamental shrub:** On Michurin's work, see Victor V. Sokolov, et al., "I. V. Michurin's Work on Expansion of the Plant Horticulture Assortment and Improvement of Food Quality," *Proceedings of the Latvian Academy of Sciences Section B* 69, no. 4 (October 2015): 190–97.

189 **Gardeners had loads of fruit:** On modern Soviet food production, see Elena Kochetkova, "Making Food Modernity: Science and Technology in Late Soviet Nutrition and Food Production," *Contemporary European History* 33, no. 2 (May 2024): 583–98.

190 **I learned why by reading:** On fermentation, see Katherine R. Amato et al., "Predigestion as an Evolutionary Impetus for Human Use of Fermented Food," *Current Anthropology* 62, no. S24 (October 2021): S207–19.

16. THE RABBIT EATERS

193 **they lacked urban infrastructure:** For information on sanitation in dachas, see Proktno-Stroitel'noe predpriiatie, "Krengol'mstroi, Proektno-Konstruktorskii otdel, Proekt cadovogo doma i khozpostroiek," Narva municipal government (no date), author's private collection.

194 **building materials were hard to come by:** For information on constructing dachas and scavenging, see Triin Ojari, "Under the Watchword of Holidays: Development of Summer Cottage Architecture During the Soviet Period," in Lankots and Ojari, eds. *Leisure Spaces*, 1–10.

194 **"just a drunk":** Author interview with Mart Pungo, Narva, Estonia, September 11, 2021.

195 **sanitizing the urban ecosystem:** Kasatkina, "Dachnye Razgovory," 87–88.

195 **access to chemical fertilizers:** Author interview with Enu Printsman, Kohtla Järve, Estonia, September 12, 2020.

196 **"I do not spray my plot":** Boukharaeva and Marloie, *Family Urban Agriculture in Russia*, 116.

196 **"then it is not food anymore":** Lilian Pungas, "Food Self-Provisioning as an Answer to the Metabolic Rift: The Case of '*Dacha* Resilience' in Estonia," *Journal of Rural Studies* 68 (May 2019): 75–86.

196 **"I was taught to observe the plants":** Author interview with Tuuli Reinso, Tallinn, September 21, 2021.

196 **"I can't eat rabbit meat":** Author interview with Tiuu Saarist.

196 **swap of carrots for Camus:** Author interview with Henn Sokk, Estonia, September 10, 2021.

197 **"part of our common wealth":** Galtz, "Strength of Small Freedoms," 183.

197 **Pensioners made extra income:** For information on income and population of Soviet garden allotments, see Markevich, "Finding Additional Income"; Anna Ivanovna, "Money, Property and Labor: Notions of Personal Wealth and Social Justice in the Soviet Union after Stalin, 1956–1991" (PhD diss., Harvard University Graduate School of Arts and Sciences), chapter 4.

198 **"called after my dad's mother":** Email correspondence with Anu Printsman, September 19, 2021.

198 **"have their hands in the ground":** Boukharaeva and Marloie, *Family Urban Agriculture in Russia*, 119.

199 **Gorbachev turned to gardens:** For more on Gorbachev's support of garden allotments, see Hedlund, *Private Agriculture in the Soviet Union*, 55.

200 **"zone of risky agriculture":** Zinaida Vasilyeva, "From Skills to Selves: Recycling 'Soviet DIY' in Post-Soviet Russia" (doctorante en sciences humaines et sociales—ethnologie Institut d'Ethnologie, Université de Neuchâtel, 2019), 2–10.

200 **read up on herbal remedies:** Author interview Tuuli Reinso, Tallinn, Sept 21, 2021.

201 **Shame is a sentiment that runs deep:** On Soviet shame, see Alexey Golubev, "The Western Observer and the Western Gaze in the Affective Management of Soviet Subjectivity," *Russian Social Science Review* 61, no. 3–4 (2020): 251–80.

202 **"Everything grew by itself!":** Mincyte, "Everyday Environmentalism."

202 **Gardeners' share of agricultural production doubled:** For information on increased garden production, see Boukharaeva and Marloie, *Family Urban Agriculture in Russia*, 123–125; and Southworth, "The Dacha Debate."

203 **urban gardens in Cuba:** For information on urban gardening in Cuba,

see Mario Gonzalez Novo and Catherine Murphy, "Urban Ag in the City of Havana: A Popular Response to a Crises," in Nico Bakker et al., eds., *Growing Cities, Growing Food: Urban Agriculture on the Policy Agenda, a Reader on Urban Ag* (1999), 339.

17. FLOWERS

207 **one of the most economically successful and sustainable:** Agronomist G. I. Shmelev argued in the 1970s that private agriculture was an "integrated feature of socialist agriculture." Katja Bruisch, "The Soviet Village Revisited: Household Farming and the Changing Image of Socialism in the Late Soviet Period," *Cahiers du Monde Russe* 57, no. 1 (2016), 81–100.

18. TERRIBLE FARMERS

212 **Charlotte, who runs the school garden:** All information on Gro-op farmers from author interviews with Gro-op farmers, June 4–8, 2021.

213 **"US navy submarine sailor":** Email correspondence with Vincent Owens, June 10, 2021.

213 **bury their sorrows in long rows:** On gardens as a vehicle for food justice, see Laura B. Delind, "Coming to Terms with Urban Agriculture: A Self-Critique," in Nuno Domingos, José Manuel Sobral, and Harry G. West, eds., *Food Between the Country and the City: Ethnographies of a Changing Global Foodscape* (Bloomsbury, 2014), 91–210.

215 **"We didn't know we were a food desert":** Author interview with Deanna West-Torrence, April 31, 2024.

216 **Kip's vision involved:** For Kip Curtis's vision for urban farming, see Kent (Kip) Curtis, and Grace Hand, "First You Need the Farmers: The Microfarm System as a Critical Intervention in the Alternative Food Movement," *Journal of Agriculture, Food Systems, and Community Development* 13, no. 2 (March 2024), 175–92; and author interview with Kip Curtis, June 4, 2021.

216 **The Anabaptist orders:** For Anabaptist farming practices, see Simon M. Evans, "Hutterite Agriculture in Alberta: The Contribution of an Ethnic Isolate," *Agricultural History* 93, no. 4 (2019): 657–81.

217 **high-priced real estate:** Email correspondence with Tim Hicks, June 7, 2021.

219 **federal and state budgets have kept farmers afloat:** On federal funding for commercial farming, see Damian Carrington, "$1m a Minute: The Farming Subsidies Destroying the World," *Guardian*, September 16, 2019.

19. THE YELLOW FOOD REVOLUTION

221 **Alex met us at the gate:** Alex is a pseudonym.

222 **Rudy strung . . . Sam cut:** These are pseudonyms for the RICI inmates.

225 **Trying to digest such a diet:** On methane emissions from livestock, see Diane Mayerfeld, Will Fulwider, and Alli Parrish, "Methane Emissions from Livestock and Climate Change," Crops and Soils Division of Extension, University of Wisconsin-Madison.

225 **The new hybrid seeds:** On the "Green Revolution," see Deborah Fitzgerald, "Exporting American Agriculture: The Rockefeller Foundation in Mexico, 1943–53," *Social Studies of Science* 16, no. 3 (1986): 457–83.

226 **"Our primary concern":** R. Douglas Hurt, *The Green Revolution in the Global South: Science, Politics, and Unintended Consequences* (University of Alabama Press, 2020), 43.

226 **He didn't do this work alone:** On Borlaug's overlooked colleagues, see Gabriela Soto Laveaga, "Beyond Borlaug's Shadow: Octavio Paz, Indian Farmers, and the Challenge of Narrating the Green Revolution," *Agricultural History* 95, no. 4 (2021): 576–608.

226 **Since the start of global enclosures:** On famines from global enclosures, see Mike Davis, *Late Victorian Holocausts: El Niño Famines and the Making of the Third World* (Verso, 2001).

226 **They say the modern food system:** On Rockefeller Foundation and failures of modern food systems, see Joseph Peralta, "Doing More with Less: Fixing Our Food System," *The Rockefeller Foundation* (blog), September 4, 2018.

227 **makes for a fragile system:** On the fragility of industrial agriculture, see Wallace-Wells, "Food as You Know It."

227 **one in four on Earth, suffered:** On malnourishment, see Bryan Chong et al., "Trends and Predictions of Malnutrition and Obesity in 204 Countries and Territories: An Analysis of the Global Burden of Disease Study 2019," *eClinicalMedicine* 57 (March 2023).

227 **Louis Bromfield bought a farm:** On Louis Bromfield in Ohio, see Stephen Heyman, *The Planter of Modern Life: Louis Bromfield and the Seeds of a Food Revolution* (W. W. Norton, 2020), 135, 169, 182.

20. *P* IS FOR PLANT

233 **vast tracts of boggy peatland:** On peatlands in the Netherlands, see J. W. de. Zeeuw, "Peat and the Dutch Golden Age: The Historical Meaning of Energy-Attainability," *AAG Bijdragen* 21 (1978): 3–31.

234 **purchasing wheat and rye from Poland:** On outsourcing food production in the Netherlands, see Moore, "Capitalocene, Part I."

234 **The tubers cross each other:** On potato imports and exports in the Netherlands, see USDA Foreign Agricultural Service, "Gain Report: Netherlands," N. NL8032, July 26, 2018; and OEC Profiles, Potatoes in the Netherlands, 2022.

234 **feed not people, but animals:** On Dutch feedstock, see Rob Roggema, "From Food as Commodity to Food as Community," in Rob Roggema, ed., *The Coming of Age of Urban Agriculture* (Springer, 2023), 1–6.

235 **nitrogen has degraded soil, water, and air quality:** On nitrogen pollution in the Netherlands, see Hans J. M. van Grinsven et al., "Benchmarking Eco-Efficiency and Footprints of Dutch Agriculture in European Context and Implications for Policies for Climate and Environment," *Frontiers in Sustainable Food Systems* 3, no. 13 (March 2019).

236 **"There is no climate crisis":** Claire Moses, "Dairy Farmers in the Netherlands Are Up in Arms over Emission Cuts," *New York Times*, August 20, 2022.

236 **"More Farmers, Less Bullshit":** see "'We willen meer boeren, minder bullshit,'" *Berkelbode*, April 29, 2024, 1.

236 **at the expense of vegetables, fruits, and nuts:** On EU agricultural subsidies, see Anniek J. Kortleve et al., "Over 80% of the European Union's Common Agricultural Policy Supports Emissions-Intensive Animal Products," *Nature Food* 5, no. 4 (April 2024): 288–92.

237 **global demand for meat is on the rise:** On meat consumption and livestock feed, see Arnold van Huis and Laura Gasco, "Insects as Feed for Livestock Production," *Science* 379, no. 6628 (January 2023).

21. GLOBALISM UNDER GLASS

247 **Most community gardens charge:** On greenhouse allotments in Almere, see Nell Westerlaken, "Bananenbomen in Almere," *de Volkskrant*, August 16, 2023.

251 **visit with their plants and neighbors:** For time-budget survey of gardeners in Soviet Russia, see Markevich, "Finding Additional Income."

251 **in most categories of crops, urban agriculture was similar to:** For information on crop yields of urban agriculture, see Florian Thomas Payen et al., "How Much Food Can We Grow in Urban Areas? Food Production and Crop Yields of Urban Agriculture: A Meta-Analysis," *Earth's Future* 10, no. 8 (2022); and Robert McDougall et al., "Small-Scale Urban Agriculture Results in High Yields But Requires Judicious Management of Inputs to Achieve Sustainability," *PNAS* 116, no. 1 (2018): 129–134.

252 **Another study found:** For a comparison of CO_2 in urban and conventional agriculture, see Jason K. Hawes et al., "Comparing the Carbon Footprints of Urban and Conventional Agriculture," *Nature Cities* 1 (2024), 1–10.

252 **about a third of food demand:** For information on food demand and urban agriculture, see Khadija Benis and Paulo Ferrão, "Potential Mitigation of the Environmental Impacts of Food Systems Through Urban and Peri-Urban Agriculture (UPA)—A Life Cycle Assessment Approach," *Journal of Cleaner Production* 140 (2017): 784–95.

254 **"community and nature were suffering":** Author interview with Wouter van Eck, Kettlebroek, Netherlands, November 23, 2023.

256 **"which I call intuitive laziness":** Author interview with Wouter van Eck, November 23, 2023.

258 **Scientists generally agree:** Josefine Lærke Skrøder Nytofte and Christian Bugge Henriksen, "Sustainable Food Production in a Temperate Climate—A Case Study Analysis of the Nutritional Yield in a Peri-Urban Food Forest," *Urban Forestry and Urban Greening* 45 (2019), 126326.

CONCLUSION: THE DIGGERS

261 **"saying This is ours":** Marks, *A False Tree of Liberty*, 88.

261 **"trade of darkness":** Marks, *A False Tree of Liberty*, 88.

262 **most productive agriculture in recorded human history:** For more on this, see Stanhill, "An Urban Agro-Ecosystem."

264 **"next renaissance of human culture":** Thomas Rainer and Claudia West, *Planting in a Post-Wild World: Designing Plant Communities for Resilient Landscapes* (Timber Press, 2015), 255.

264 **"polyfood crisis":** Wallace-Wells, "Food as You Know It."

264 **fresh food grown in organic soils:** For information on the nutritional benefits of organic gardens, see Elizabeth G. Berg, "Bringing Food Back Home: Revitalizing the Postindustrial American City Through State and Local Policies Promoting Urban Agriculture," *Oregon Law Review* 783 (2014); Parajuli et al., "Yard Vegetation Is Associated"; Gavin Pereira et al., "The Association Between Neighborhood Greenness and Weight Status: An Observational Study in Perth Western Australia," *Environmental Health* 12, no. 1 (June 2013): 49; and G. Pereira et al., "The Association Between Neighborhood Greenness and Cardiovascular Disease: An Observational Study," *BMC Public Health* 12, no. 1 (2012): 466.

265 **"rural mortality penalty":** Wesley L. James et al., "Conceptualizing Rurality: The Impact of Definitions on the Rural Mortality Penalty," *Frontiers in Public Health* 10 (2022).

Image Credits

Various design elements: Ilyanna Kerr

xii Official British photo from OWI
57 Heinrich Zille
69 Landesarchiv Berlin, F Rep. 290 (09) Nr. 0295261 / Foto: k. A.
69 Landesarchiv Berlin, F Rep. 290 (09) Nr. 0017472 / Foto: Cürlis, Peter
87 Landesarchiv Berlin, F Rep. 290 (09) Nr. 0246119 / Foto: k. A.
95 Harris & Ewing
100 John Wymer
124 Landesarchiv Berlin, F Rep. 290 (09) Nr. 0017472 / Foto: Cürlis, Peter
140 John Wymer
143 John Wymer
152 Georgette Norman
155 Georgette Norman
183 Author photo
223 Author photo
238 Author photo
242 Author photo

Index